路易斯·I.康：
光与空间

[瑞士] 乌尔斯·布提克 | 著
Urs Büttiker

卢紫荫 | 译

Louis I.Kahn:
Light and Space

天津大学出版社
TIANJIN UNIVERSITY PRESS

Louis I.Kahn: Light and Space
Copyright © by Urs Büttiker
Chinese translation copyright © 2019 by Tianjin University Press
All rights reserved

版权合同：天津市版权局著作权合同登记图字第 02-2017-62 号
本书中文简体字版由 Urs Büttiker 授权天津大学出版社独家出版。

路易斯·I. 康：光与空间 | LOUIS I.KAHN: GUANG YU KONGJIAN

图书在版编目（CIP）数据

路易斯·I. 康：光与空间 /（瑞士）乌尔斯·布提克著；卢紫荫译. -- 天津：天津大学出版社，2019.5（2023.6 重印）
ISBN 978-7-5618-6397-8

Ⅰ. ①路… Ⅱ. ①乌… ②卢… Ⅲ. ①康 (Kahn, Louis Isadore 1901-1974) – 建筑艺术 – 艺术评论 Ⅳ. ① TU-867.12

中国版本图书馆 CIP 数据核字 (2019) 第 086915 号

出版发行	天津大学出版社
地　　址	天津市卫津路 92 号天津大学内（邮编：300072）
电　　话	发行部：022-27403647
网　　址	www.tjupress.com.cn
印　　刷	北京华联印刷有限公司
经　　销	全国各地新华书店
开　　本	235mm×220mm
印　　张	16.5
字　　数	130 千
版　　次	2019 年 5 月第 1 版
印　　次	2023 年 6 月第 3 次
定　　价	69.00 元

凡购本书，如有缺页、倒页、脱页等质量问题，烦请与我社发行部门联系调换。

版权所有　　侵权必究

谨以此书献给埃斯特·康夫人
THIS BOOK IS DEDICATED TO
MRS. ESTHER KAHN

路易斯·I.康创作的木版画。本图是根据1931年埃斯特和路易斯·I.康制作的圣诞贺卡复制。

Woodcut created by Louis I. Kahn. Reproduction of Christmas card used by Esther and Louis I. Kahn in 1931.

目录
CONTENTS

10	引言		64	宾夕法尼亚太阳屋
16	**INTRODUCTION**			PENNSYLVANIA SOLAR HOUSE
22	光与空间		66	菲利浦·Q. 罗奇住宅
30	**LIGHT AND SPACE**			PHILLIP Q. ROCHE HOUSE
40	学生时代作品 购物中心 STUDENT WORK SHOPPING CENTER		68	莫顿·韦斯住宅 MORTON WEISS HOUSE
			72	费城精神病医院伯纳德·S. 平克斯楼 BERNARD S. PINCUS BUILDING, PHILADELPHIA PSYCHIATRIC HOSPITAL
42	泽西住宅项目 JERSEY HOMESTEADS		74	费城精神病医院塞缪尔·拉德贝尔楼 SAMUEL RADBILL BUILDING, PHILADELPHIA PSYCHIATRIC HOSPITAL
44	艾哈瓦以色列犹太教会堂 AHAVATH ISRAEL SYNAGOGUE			
46	预制住宅研究 PREFABRICATED HOUSE STUDIES		76	耶鲁大学美术馆 YALE UNIVERSITY ART GALLERY
48	杰西·奥泽住宅 JESSE OSER HOUSE		80	伦纳德·弗鲁赫特住宅 LEONARD FRUCHTER HOUSE
50	L. 布鲁多住宅 L. BROUDO HOUSE		82	美国劳工联合会医疗服务楼 AMERICAN FEDERATION OF LABOR MEDICAL SERVICES BUILDING
52	卡佛庭院住宅开发项目 CARVER COURT HOUSING DEVELOPMENT		84	犹太社区中心 JEWISH COMMUNITY CENTER
54	194X 旅馆 HOTEL FOR 194X		86	劳伦斯·莫里斯住宅 LAWRENCE MORRIS HOUSE
56	国际女装工人工会健康中心 INTERNATIONAL LADIES GARMENT WORKERS UNION HEALTH CENTER		88	华盛顿大学图书馆竞赛方案 WASHINGTON UNIVERSITY LIBRARY COMPETITION
58	阳伞住宅 PARASOL HOUSES		90	伊莱恩·考克斯·克莱弗住宅 ELAINE COX CLEVER HOUSE
60	费城精神病医院 PHILADELPHIA PSYCHIATRIC HOSPITAL		92	阿尔弗雷德·牛顿·理查德医学和生物研究中心 ALFRED NEWTON RICHARDS MEDICAL RESEARCH BUILDING AND BIOLOGY BUILDING
62	B.A. 伯纳德住宅 B.A. BERNARD HOUSE		94	论坛报出版公司大楼 TRIBUNE REVIEW PUBLISHING COMPANY BUILDING

96	罗伯特·H. 弗莱舍住宅 ROBERT H. FLEISHER HOUSE	148	贝塞尔犹太教会堂 TEMPLE BETH-EL SYNAGOGUE
98	玛格丽特·埃西里科住宅 MARGARET ESHERICK HOUSE	150	胡瓦犹太教堂 HURVA SYNAGOGUE
104	第一唯一神教堂及学校 FIRST UNITARIAN CHURCH AND SCHOOL	154	沃尔夫森机械与运输工程中心 WOLFSON CENTER FOR MECHANICAL AND TRANSPORTATION ENGINEERING
108	美国领事馆及馆舍住宅 U.S. CONSULATE AND RESIDENCE	156 157	耶鲁英国艺术中心 YALE CENTER FOR BRITISH ART
110	索克生物研究所 SALK INSTITUTE FOR BIOLOGICAL STUDIES	162	计划生育与妇幼保健中心 FAMILY PLANNING CENTER AND MATERNAL HEALTH CENTER
114	表演艺术中心 PERFORMING ARTS CENTER	164	联合神学研究院公共图书馆 COMMON LIBRARY, GRADUATE THEOLOGICAL UNION (GTU)
118	埃莉诺·唐纳利·厄德曼礼堂 ELEANOR DONNELLY ERDMAN HALL	166	孟加拉首都政府建筑群：医院 SHER-E-BANGLA NAGAR: NATIONAL CAPITAL OF BANGLADESH HOSPITAL
120	费城艺术学院 PHILADELPHIA COLLEGE OF ART	168 169	孟加拉首都政府建筑群：议会堂 SHER-E-BANGLA NAGAR: NATIONAL CAPITAL OF BANGLADESH ASSEMBLY HALL
122	诺曼·费舍住宅 NORMAN FISHER HOUSE		
124	米克韦以色列犹太教会堂 MIKVEH ISRAEL SYNAGOGUE	186 187	**光线控制类型** **TYPOLOGY OF LIGHT CONTROL**
128	印度管理学院 INDIAN INSTITUTE OF MANAGEMENT	190	**参考文献** **REFERENCES**
134	美国风交响乐团驳船 BARGE FOR THE AMERICAN WIND SYMPHONY ORCHESTRA	191	**图片来源** **FIGURE CREDITS**
136	菲利普斯·埃克塞特学院图书馆 PHILLIPS EXETER ACADEMY, LIBRARY	192 193	**第一版致谢** **CREDITS FOR THE FIRST EDITION**
140	奥利维蒂 – 安德伍德工厂 OLIVETTI-UNDERWOOD FACTORY	194 195	**中 / 英文对照版致谢** **CREDITS FOR THE CHINESE/ENGLISH EDITION**
142	金贝尔美术馆 KIMBELL ART MUSEUM		

引言

路易斯·I. 康与他的时代

"结构设计实际就是光的设计。拱顶、穹隆、拱、柱子皆是呼应光的各种特质的结构体。自然光在一年四季、一日四时中的微妙变化，营造出不同的空间氛围，犹如它进入了空间，并且修饰着空间。"（1）

路易斯·艾瑟铎·康（Louis Isadore Kahn）1901年生于爱沙尼亚。四岁时随父母移民费城。艺术天赋使他得以进入宾夕法尼亚大学建筑系学习。他在校学习建筑期间以及随后跟随保罗·菲利浦·克瑞（Paul Philippe Cret）教授的学习生涯，都完全遵从传统的鲍扎体系原则。

"……宾大是一所好学校。这里校风严谨，它并非一个门派，而是一种值得尊重的卓越品质……在这里，我们学会去尊重大师的作品，并非他们为自己所做的事情，而是他们通过作品为其他人所做的一切，并且用大量的建筑语言表达出来。"（2）

取得学位后，康曾在几个建筑工作室工作过，包括市政建筑师约翰·莫利托和他从前的教授保罗·菲利浦·克瑞的工作室。1928年，康回到他的出生地爱沙尼亚旅行，之后又游历了英国、荷兰、德国、意大利和法国。尽管他有机会参观了在现代主义影响下建成的许多建筑，但他更喜欢去拜访著名的古迹遗址。康与现代建筑的邂逅是来自现代主义追随者们的出版物。当被问及关于现代建筑的著作和建筑作品时，他这样回答：

"当我初次认识现代建筑时，相关书籍炙手可热。我非常喜欢那些书，但看起来怎么都不及勒·柯布西耶的考察研究、他的草图、他的思考、他的观点那样耀眼。我还认识到，一个我在宾大所受到的训练中从未考虑过的简单住宅，也可以变得非常重要。"（3）

你可以想象这样的场景：当勒·柯布西耶建造萨伏伊别墅时，路易斯·I. 康正在他的教授保罗·菲利浦·克瑞的工作室工作，设计着福尔杰·莎士比亚图书馆。这两种建筑语言的差异非常大。因此康与同时代的建筑师们不可避免地要决定到底选择哪一种建筑观点。至少，乔治·豪（George Howe）和威廉·莱斯卡兹（William Lascaze）设计的费城储蓄基金会大楼标志着现代主义的影响在康的家乡已成为不可忽视的力量。他也不再对此无动于衷。现代建筑原则自然而然地影响了他的设计理念，在那之前，他受到的完全是鲍扎体系观念的教育。

"勒·柯布西耶对我有很大的启示。他让我认识到一个活着的人也能够成为精神导师。他用他的作品来教导你……我觉得可以扔掉那些书本了，那些当时沃尔特·格罗皮乌斯（Walter Gropius）倡导要扔掉的书本……"（4）

即使是1932年的大萧条时期，也没有阻止康与"现代建筑原则"的交锋。他与其他30位失业的建筑师和工程师共同创办了建筑研究小组（1932—1934）。

在接下来的几年中，康的设计涉及一些公众支持的住宅项目和城市设计。理想城市规划创作于1937—1939年，是为容纳两百万居民的费城设计的城市愿景。然而，那时这个项目并没有路易斯·I. 康的个人特征，而是1922年勒·柯布西耶的"光辉城市"的一个改良版。

1935—1937年间，阿尔弗雷德·卡斯特纳与路易斯·I. 康共同合作规划建造了位于新泽西州罗斯福的泽西住宅项目。这个项目包括大约250个居住单元，所有建筑完全遵循现代主义建筑运动的理念。在同一时期，康第一次有机会完全独立完成了一个项目：艾哈瓦以色列犹太教会堂。这幢建筑是一个纯净的立方体，在一排传统的住宅之间，因其十分简洁的体量而凸显出来。内部却出人意料地令人感到温暖与高贵。1924年获得建筑学位后，康实际建成的项目仅有这两个。

这种境遇随着第二次世界大战的爆发发生了改变。康在与乔治·豪和奥斯卡·斯托罗诺夫（Oscar Stonorov）合作时，仅在1941—1942年间，就至少设计了7个战争住宅项目。其中的5个实际建成。康终于有机会在大尺度上去实践并检验现代建筑理念。

1945年，康完成了对B.A. 伯纳德住宅的加建。这个加建项目多少有些失败，似乎标志着康的作品中一个重要的休整期。康对这块问题区域没有采用熟练的处理方式：几何形的柱子、带形窗、平屋顶、一个像是悬浮在厚重基座上的立方体、垂直的滑动墙板，等等。康在韦斯住宅（1947—1950）中再次证明了他对建筑的掌控能力。这个项目成功之后，随之而来的是犹太社区中心（1948—1954）以及康的第一个建成代表作——耶鲁大学美术馆（1951—1953）。在这个时期，康真正找到了自己独特、原创的建筑语言，在接下来的设计生涯中，他以非凡的创造力对它进行了发展。

对一位建筑师来说，很难如康一样，在接受鲍扎体系理念如此深入的教育之后，能快速而轻松地卸下从前的一切，转而无条件地支持现代建筑。事实上，对康来说，也经历了相当漫长而乏味的过程，将现代的方式与以往已被验证的方式、将平凡与高尚整合在一起，从而探索出自己独特的设计方法。

"你永远无法习得不属于你的东西。你所习得的一切，都是与你密切相关的，只是在你身上没有真实体现出来……"（5）

1947年，康开始在耶鲁大学演讲，8年后，他开始在宾夕法尼亚大学任教。

1950—1951年间，康应邀到位于罗马的美国建筑师学院做驻校建筑师，此间的学习机会对康的设计发展起到了至关重要的作用。他利用驻外的机会游历了意大利、埃及和希腊，研学了那里的建筑遗产。吉萨金字塔、卡纳克神庙的巨大柱子、卢克索神庙强有力的大门、雅典卫城以及卡拉卡拉浴场的砖拱，激发了康对建筑世界的思索与感受。

"希腊建筑让我明白，柱子是没有光的地方，而柱子之间是有光的地方。一列柱子就是无光、有光、无光、有光的变化。柱与柱之间让光线射进来。由墙衍生出的柱，形成了自己的韵律，无光、有光、无光、有光，这是艺术家的奇迹。"（6）

之后，康的作品越来越多地呈现出新颖又复古的特质，使得他在同时代建筑师中越来越特立独行。他对设计问题的本质或本源的追索引领他在建筑历史中探寻。在人类准备进入外太空的历史时刻，他从埃及古迹、希腊神庙、罗马浴场、苏格兰城堡、法国城市的防御工事、哥特大教堂、文艺复兴教堂以及意大利广场中获得灵感。但这并不表示康的建筑简单地复制了某个历史时代；相反，他致力于探索建筑问题解决方法的普遍本质，然后尝试用现代技术将其实现。同时，路易斯·I.康充分发展了他的建筑语言，也形成了自己独有的语言表达方式。他开始以一种全新的、独特的方式使用词汇，并将词汇组合在一起，以便更好地表达他的观点和哲学思索。他的文字总是很难读懂，难以解释，更难以理解。如他所说："即使一个词语也是艺术创作。"（7）

康解决建筑问题的方式深奥而复杂，这反映在他的设计方法中。他的方法包括用木炭笔在草图纸上画一条线，下一刻就会把它擦掉，以期下一条线会是"更恰当的"一条。康复杂而耗时的设计方法，很大程度上与细节有关，但最终常常导致项目不能在约定时间内完成。而且，他也很难接受委托人的空间规划并在成本控制下进行设计。康的许多设计从一开始就规模过大，成本也太高。重新规划或是从头再来常常是不可避免的。最终，康被疏远，并失去了许多客户。有时候，尽管开发商已经与他解约，他也会继续设计：仅仅是因为他无法舍弃。还有许多项目康做了许多年，也没能得到一个满意的结果。更不用说，这种设计方法也不一定经济；事实上，路易斯·I. 康死后，他的工作室已负债累累。他将自己的生命与努力全部献给了建筑，已无法想象以任何其他方式生活。

一次，他从印度和孟加拉国的项目归来，直接从机场回到工作室，此时是凌晨5点钟。他发现一位雇员仍在绘图室通宵工作。他惊讶地问："其他人呢？"

"没有什么像建筑一样。建筑精神存在，却无形。唯有建筑作品是有形的。最好是将建筑作品看作为建筑本身所奉献的祭品，只是因为始于最初的惊叹。"（8）

笔记
路易斯·I. 康

Notices
Louis I. Kahn

My remark that structure is the giver of light is thereby recalled in the march of columns in the Greek Temple.

light no light .

no light light no light no light.

The column is like the since of the pen where the light is not.

INTRODUCTION

LOUIS I. KAHN AND HIS AGE

"The structure is a design in light. The vault, the dome, the arch, the column are structures related to the character of light. Natural light gives mood to space by the nuances of light in the time of the day and the seasons of the year, as it enters and modifies the space."(1)

Louis I. Kahn was born in Estonia in 1901. When he was four years old, his parents emigrated to Philadelphia. His artistic talent made it possible for him to attend the Department of Architecture of the University of Pennsylvania. The approach to architecture at the school, then under the direction of Professor Paul P. Cret, was entirely dedicated to the tradition of the Ecole des Beaux-Arts.

"... Penn was a nice school then. It was highly religious, not as if it were a certain religion, but religious in the sense that transcendent qualities were considered worthy.... We learned to respect the works of the masters, not so much for what they did for themselves, but for what they did for others through their works, which were a high use of the language of architecture." (2)

After completing his degree, Kahn worked in a number of architectural offices, including those of the municipal architect John Molitor, and of his former professor Paul P. Cret. In 1928, Kahn returned on a trip to Estonia, his birthplace, and continued on with a tour of the Continent to England, Holland, Germany, Italy, and France. Although he had the opportunity then to visit buildings constructed under the influence of Modernism, he preferred rather to visit the famous sites of antiquity. Kahn's encounter with modern architecture took place instead on the level of publications by its adherents. He replied as follows when asked about the writings and the works of modern architecture:

"When I was first introduced to modern architecture, the books were being burned. The books that I care for a great deal somehow did not look as bright as the pages of Le Corbusier's explorations, his drawings, he speculations, his points of view. I even learned that a simple house, which was never a consideration in my training at the University of Pennsylvania, became something of importance." (3)

One may try to imagine the situation: at the same time that Le Corbusier was building the Villa Savoye, Louis I. Kahn was working in the office of Paul P. Cret, his former professor, on the design for the Folger Shakespeare Library. The disparity between these two architectural languages could not have been greater. Kahn and other architects of his generation were therefore not able to avoid the inner conflict involved in making a decision for the one or the other of these attitudes toward architecture. At any rate, construction of the Savings Fund Society Building in Philadelphia by G. Howe and W. Lascaze marked the point at which the influence of Modernism could no longer be overlooked in Kahn's home town. And he did not remain impassive. The principles of modern architecture entered thereafter, almost of their own accord, into the world of his concepts until then so thoroughly permeated by the attitudes of the Ecole des Beaux-Arts.

"Le Corbusier was a revelation to me. He made me realize that there was a man alive who could be an inspiring teacher. Through his work he could teach you.... I felt I could throw away the books, which was what was in the air at the time Gropius advocated throwing away the books...." (4)

Even the Great Depression around 1932 was not able to delay Kahn's fundamental confrontation with the "rules of modern architecture." Together with thirty other unemployed architects and engineers, Kahn founded The Architectural Research Group (1932 - 1934).

During the following years, Kahn became involved with publicly supported housing projects and with urban design. The Rational City Plan was created during the years 1937 - 1939, as an urban vision for the accommodation of two million residents in Philadelphia. This project, at that point in time, however, did not carry the personal signature of Louis I. Kahn: rather, it represents a modified version of Le Corbusier's "Ville contemporaine" of 1922.

During 1935 to 1937, Alfred Kastner and Louis I. Kahn collaborated to plan and build the Jersey Homesteads in Roosevelt, New Jersey. This was a settlement of approximately 250 residential units, the architecture of which was entirely dedicated to the concepts of the Modern Movement. During the same period, Kahn had the first opportunity of executing a project entirely on his own: the Ahavath Israel Synagogue. The building, a pure cube in a conventional row of houses, is marked by great simplicity. The interior surprisingly radiates warmth and nobility. Since receiving his architectural degree in 1924, Louis I. Kahn did not have the opportunity to actually execute more than these two projects.

The situation changed with the outbreak of World War II. In collaboration with G. Howe and O. Stonorov, Kahn designed no fewer than seven War Housing Projects, during the years 1941 to 1942 alone. Five of them were actually constructed. Kahn finally had the chance to implement and to test certain ideas of modern architecture on a large scale.

In 1945 Kahn built an addition to the B.A. Bernard House. This addition created a somewhat awkward impression which apparently represents an important caesura in Kahn's work. Kahn did not provide skillful solutions for the problem areas treated here: the geometry of the columns, the strip window, the flat roof, a cube apparently hovering over a ponderous footing, the vertical clapboard siding, etc. It was only with the Weiss House (1947 - 1950) that evidence appeared which reaffirmed Kahn's architectural mastery. This success was followed by the Jewish Community Center (1948 - 1954) and by Kahn's first major work executed for the Yale University Art Gallery (1951 - 1953). By this time Kahn had indeed found his own individual and original architectural language which he would prove in the following years to develop further with extraordinary creativity.

It would hardly have been possible for an architect who had carried the attitudes and the conceptional modes of the Ecole des Beaux-Arts so deep in his own being as did Kahn, to quickly and effortlessly carry off the attempt to become an unconditional proponent of modern architecture. Indeed, with Kahn it required a very long and tedious process to integrate the modern with the tried-and-proven, and the prosaic with the noble, into a process which bore his own, individual signature.

"You can never learn anything that is not part of yourself. Everything you learn is attached, glued on, but it has no real substance in you...." (5)

Kahn began to lecture at Yale University in 1947, and eight years later he was appointed to teach at the University of Pennsylvania.

The study opportunities which Kahn enjoyed in Rome at the American Academy, as Resident Architect from 1950 to 1951, played a key role in his development. He took advantage of his foreign stay to visit Italy, Egypt, and Greece and to study the architectural legacies there. His experience with the Pyramids in Giza, the mighty columns of Karnak, the powerful pylons of Luxor, the Acropolis, and the brick vaults of the Caracalla Baths could not help but to have stimulated his thinking and feeling for the architectural world.

"Greek architecture taught me that the column is where the light is not, and the space between is where the light is. It is a matter of no-light, light, no-light, light. A column and a column brings light between them. To make a column which grows out of the wall and which makes its own rhythm of no-light, light, no-light, light: that is the marvel of the artist." (6)

Afterward, Kahn's projects increasingly demonstrated innovative and archaic characteristics, which make him more and more of a loner of his age. His search for the essence or the origin of a design problem led him on expeditions through architectural history. He gained inspiration from

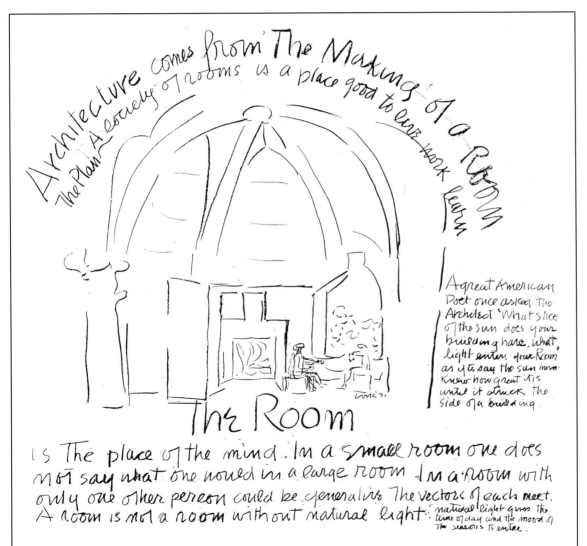

草图
路易斯·I.康

Sketch
Louis I. Kahn

Egyptian monuments, Greek temples, Roman baths, Scottish castles, fortified French cities, Gothic cathedrals, Renaissance churches, and Italian squares at a point in history when mankind prepared to enter outer space. This does not mean that Kahn's architecture simply copied the style of a particular historical epoch; rather, his efforts represented the search for the universally valid essence of a solution to an architectural problem, which he then attempted to implement through modern techniques. At the same time Louis I. Kahn fully evolved his architectural language, he also developed his own, personal form of verbal expression. He began to employ and to combine words in a new and unique manner, in order to better express his ideas and philosophical reflections. His texts are often difficult to read, to interpret, and to understand. As he stated, "Even a word is a work of art." (7)

Kahn's profound and complex manner of solving architectural problems was reflected in the methodology which he employed in his design process. His technique consisted of drawing a line with soft charcoal on sketch paper, then erasing it in the next moment in the hopes that the next line would be "the more appropriate." Kahn's complicated and time-consuming method of design, in conjunction with an intensive degree of involvement with details, had the result that he, often enough, was not able to finish a project by the agreed time. In addition, he had difficulty in accepting space assignment plans from his clients, and in working within cost limits. Many of Kahn's designs were overdimensioned from the very beginning and were much too expensive. Redrawing the plans, or even beginning again, was often unavoidable. As a result, Kahn alienated and lost many of his clients. In some cases, he continued to work on a design, although the developer had already taken the contract away from him: he simply could not part with it. There were also many projects on which Kahn worked for years on the plans without reaching a satisfactory solution. Needless to say, this working technique was not exactly economical; indeed, when Louis I. Kahn died, his office was deep in debt. He fully dedicated his life and all his efforts to architecture, and could not imagine living any other way.

Once, on returning from a project in India and Bangladesh, he went straight from the airport to his office, where he arrived at five in the morning. He found one employee still in the drafting room who had just worked through the entire night. "But where are the others?" was his astonished question.

"There is no such thing as architecture. There is the spirit architecture, but it has no presence. What does have presence is a work of architecture. At best it must be considered as an offering to architecture itself, merely because of the wonder of its beginning." (8)

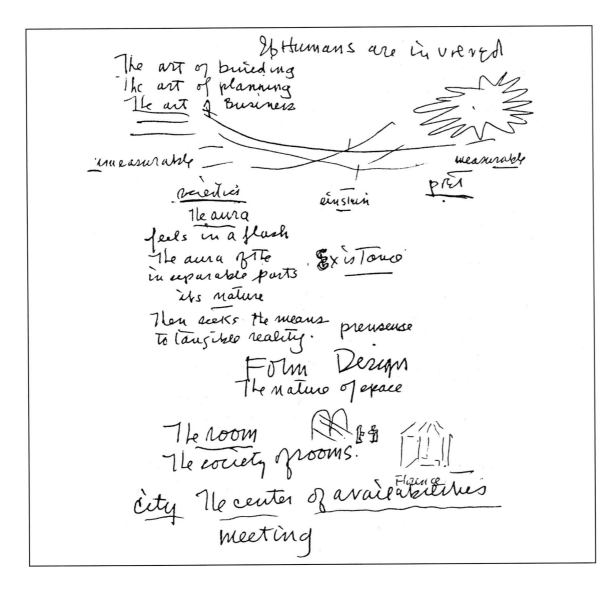

Notices
Louis I. Kahn

光与空间

本书的萌生

"学校始于一个人在一棵树下与他人讨论他的领悟,他不知道自己是老师,其他人也不知道自己是学生。学生们回忆起与这个人一起交流时的愉悦。他们想让自己的孩子也能聆听这样一个人的讲述。很快,搭建起所需的空间,第一所学校诞生了。学校的产生是必然的,因为它是人类渴望的一部分。"(9)

1976—1977 年,我在苏黎世联邦理工大学学习的第三年和第四年,有幸跟随来自提契诺的建筑师伊凡诺·吉亚诺拉和利维奥·瓦契尼学习。他们让我认识了路易斯·I. 康的建筑作品,并帮我熟悉这些作品。我第一次读到了如此诗意的叙述,诸如"建筑始于光而终于影"以及"天空是广场的屋顶",还有"没有自然光的房间就不是房间"。

这些文字铭刻在我的脑海中,给我留下了深深的印象,也激发了我的兴趣,让我去更多地了解路易斯·I. 康和他的作品。我开始大量阅读能找到的所有关于他的出版物。很快我就意识到,必须去看看他的作品才能真正理解它们。

1980 年,在纽约雪城大学任教一年后,我有机会拜访并拍摄了路易斯·I. 康在美国的几乎所有作品。学术旅行期间,我画下了看到的关于他的一切,逐渐地,我开始确定这些作品中最重要的主题:路易斯·I. 康建筑中的光。他对光的控制、他的光线调节构件、他设计的建筑中闪烁的光影、与空间和结构相关的光的数量以及在一年变幻的自然光中建筑的变化。

埃西里科住宅(1959—1961)尤其吸引我:康对光与空间的熟练掌控,是他的住宅设计中最令人印象深刻的部分。

1980 年夏天我第一次造访埃西里科住宅。那时已经确定,我会再次来到这幢住宅,更多地学习和体验。我想去探访这个似乎充满秘密的住宅。现在回想起来,终于知道为何康的创作如此吸引我:因为变幻的光和一年中光的变化,使这幢住宅每次拜访都呈现出不同的特质。

后来我决定回到费城,到宾夕法尼亚大学利用康的档案馆去研习康的毕生作品。我想通过原始草图、照片、信件和他早期合作者的资料对其作品更加熟悉。

我最初的计划只是想要整理埃西里科住宅的详尽资料，因为它在康的所有作品中非常重要，但出于一个很好的理由，这个计划很快就被放弃了。在翻阅了关键词"埃西里科住宅"下的原始草图和微缩胶片后，我惊讶地发现，现在的埃西里科住宅原本还有一个加建的部分，而且还有一个为艺术家沃顿·埃西里科建成的工作室。这两个项目怎么可能直到康去世10年后一直没有被发现，尽管此间还出版有一本名为《路易斯·I. 康：1935—1974作品全集》的书？

更巧合的是，这个发现促使我系统地检索了康的作品的所有25000个微缩胶片档案。我坐在"微缩阅读复印机"的屏幕前，发现了大量康先前没有出版过的作品。我复制了这些微缩胶片文件，收集了几乎所有由康设计和建成的项目档案。但我该如何分析与进一步使用这个巨大的新发现呢？

最终，我决定扩展自己初始的想法，即不止研究埃西里科住宅的光线控制、窗户设计、可调节的内外视野、整合的通风嵌板、遮阳板及其他构件。而是将研究对象拓宽到一些选定的设计，很快也确定了我的研究主题："路易斯·I. 康作品中的光与空间"。我感觉这个清晰确定的研究背景能让我系统地对康的项目进行比较研究。

我向康早期的同事了解情况后，带着一张城市地图，坚持寻找到当时康还不为人知的那些作品。我拍下照片，并尽我所能去了解它们。

1982—1983年间，我整理出24个文档，在分析草图的帮助下，厘清了路易斯·I. 康建筑中光的核心主题。我以复制的微缩胶片文件作为这项个人研究的基础。叠覆技术的使用揭示了关于光的主题的建筑元素。用这种方式形成的对每个建筑的简介是我所做的摘要，是个人的理解和诠释。

除了草图和拼贴画，我的工作还形成了一段题为《窗户的一天》的视频。借助于视频技术，我展示了埃西里科住宅西侧起居室的窗户每隔30分钟的变化，试图记录下连续24小时内这个窗户营造的光影变化。

1985年，我开启了第三次的路易斯·I. 康考察之旅，这次的目的地是印度、尼泊尔和孟加拉国。

探访孟加拉国首都政府建筑群并拍照是我所有旅行中最困难的任务。我在政府办公室等候相关的官员等了好多天，利用茶余饭后的时间努力向他们解释我想做的事情。两周后，我终于找到了相关负责的官员，获准在国会大厦中一些通常对外禁入的相对安静的房间中拍照。

达卡的项目是康最激动人心、也是最复杂的作品，这更使这个精彩绝伦的综合建筑中许多独特的、宏伟的房间至今仍不为公众所知。

直到第一次进入议会堂的一刻，我才知道康如何处理议会堂顶部的光线（1983年时没能获得许可查看结构图纸）。因此当我终于站在这个巨大的30米高的厅堂中时，感到很惊讶也很激动，我注视着阳光如何经过外部环形结构的反射调节透过半圆形的洞口，射入议会堂。

本书最后的一些资料是我在1991年冬游历特拉维夫及1992年秋游历费城时收集的。

我从康的作品中选入本书的50个项目展示了康为了控制和调节自然光所采用的各种技术。对于每个特定的建筑，他都尝试去找到一种独特的自然光解决方式。康将窗户作为一种能够对光线进行微调的工具，经过设计可以在变化常新的条件下在外部与内部之间有效地构建一个交界面。

"没有自然光，建筑中的空间就不能成为场所。人工光是夜晚固定位置的吊灯发出的光，无法与变幻莫测地闪烁的自然光相提并论……"（10）

各种不同的自然光控制方法以及不同的光线调节元素，可以作为贯穿康设计始终的主题进行比较研究。在分析过程中，我按照时间顺序研究这些项目，从1924年康学生时代的作品开始，到1974年的达卡议会堂。我自始至终的目的是揭示并记录下康对于控制落入房间中光线的新方法的不懈追求。

窗户（window）一词来自古词汇的"wind"（风）和"eye"（眼睛），大量不同的窗户设计形式是我研究的焦点：窗是墙体或屋顶上的洞口，光线（和空气）穿过窗口进入房间。路易斯·I.康的整个职业生涯都与这个照明通风构件有关。他将每个窗户看作一个正式的功能实体，反

分析草图
乌尔斯·布提克 1983

Analytical sketches
U.B. 1983

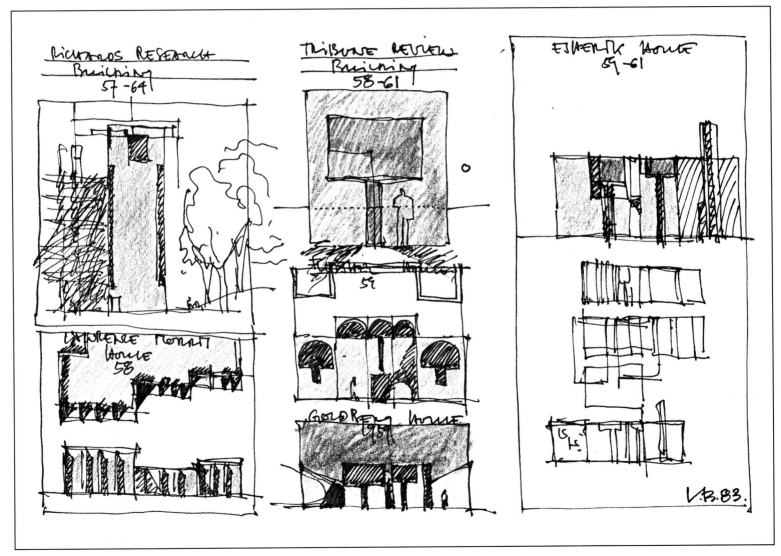

过来，这扇窗也成为他的空间结构中非常重要的组成部分。

"建筑的设计图应被读解为阳光中空间的协奏曲。即使是一个黑暗的房间也应当在神秘的洞口处有足够的光线，我们才能知道它到底有多暗。每个空间都是由其中自然光的结构和特质来定义的。"（11）

在早期作品中，康基本上采用的是传统的解决方法，其表现形式通常是独立、突出的遮阳构件。然而在晚期，他发明了许多整体的遮阳结构，赋予了他的建筑独特而强烈的个人特质。康的许多尝试还包括窗和通风口的功能和形式的整合或拆分。

为了适应安哥拉、印度、巴基斯坦、孟加拉及以色列极端的气候条件，阳光必须经过过滤，由此产生的复古洞口和形式使康的建筑拥有了一种独特的气质。他设计了巨大的"光过滤器"，一方面能够遮阳，另一方面可以保证有效的通风和望向外面的视野。

"我在印度、巴基斯坦以及现在的孟加拉国设计的建筑，都非常尊重光的秩序，建筑的外部在阳光之下，内部则成为居所。"（12）

在路易斯·I.康的晚期作品中能清晰地看到，天窗变得越来越重要，康对光的控制和调节也越来越精到。康的所有尝试无疑在达卡议会堂达到顶点。这里，我们清楚地看到康将早期设计中的经验整合起来，创造了一个真正的建筑宝藏。

这本书的组织与我自己的工作过程一致。最初的草图（和埃西里科住宅的草图）只是想展示一个工作过程，在绘制那些案例分析草图的过程中，我明确了关于光的主题并逐步发展。

本书展示了 1982—1983 年间 24 份资料的内容。这些资料（规划图和草图）在书稿的左页，我自己的解读则在对页：右上角的小图示是对"光与空间"主题的抽象概念图。第 39 页诠释了对原始设计图、叠覆草图、照片及图示和符号的分析方法。

针对埃西里科住宅之后的项目，书稿将更多的空间专门留给照片，不再提供细部草图分析。

分析草图
乌尔斯·布提克 1983

Analytical sketches
U.B. 1983

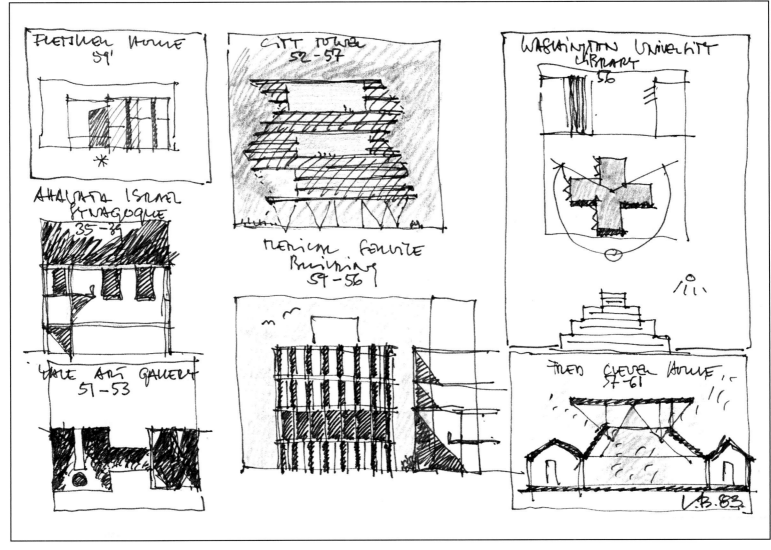

本书的最后有康设计的各种光的控制和调节构件的类型图。

"物质伴光而生,"康写道:"人是消耗的光,山脉是消耗的光,树木是消耗的光,空气是消耗的光。所有物质都是消耗的光。"(13)

分析草图
乌尔斯·布提克 1983

Analytical sketches
U.B. 1983

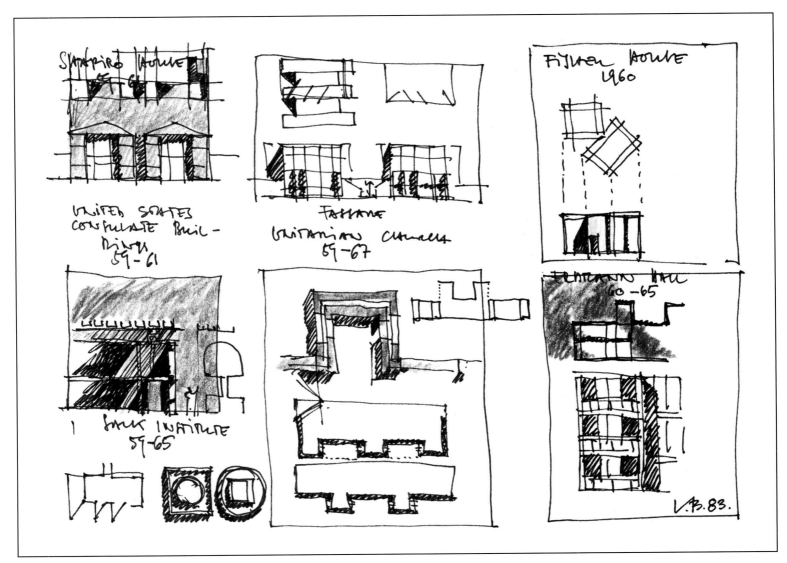

LIGHT AND SPACE

HOW THIS BOOK CAME ABOUT

"Schools began with a man under a tree, a man who did not know he was a teacher, discussing his realizations with a few others who did not know they were students. The students reflected on the exchanges between them and on how good it was to be in the presence of this man. They wished their sons, also, to listen to such a man. Soon, the needed spaces were erected and the first schools came into existence. The establishment of schools was inevitable because they are part of the desires of man." (9)

In 1976 and 1977, during the third and fourth years of my studies at the Swiss Federal Institute of Technology in Zürich, I had the privilege to learn from the architects Ivano Gianola and Livio Vacchini, both from Tessin. They pointed out to me the work of Louis I. Kahn and helped me to become acquainted with it. For the first time, I read such poetical statements as "A building begins with light and ends with shadows," and "The sky is the roof of a square," and "A room without natural light is not a room."

Such quotations as these stuck in my mind. They left a deep impression and stimulated my interest to learn more about the man Louis I. Kahn and his work. I began to devour all the publications of his that I could get my hands on. It didn't take long to realize that I would have to see his works themselves to really understand them.

In 1980, after spending a year teaching at Syracuse University in New York, I had the opportunity of visiting and of photographing almost all the works of Louis I. Kahn in the United States. During my study trip I sketched everything of his that I saw, and step by step I began to firm up the topic which appeared most essential in these works: light in the architecture of Louis I. Kahn. His control of light, light modulation, the play of light and shadow in his buildings, the amount of light in relationship to space and structure, and the change which a building undergoes as natural light varies throughout the year.

The Esherick House (1959 – 1961) especially fascinated me: Kahn's virtuoso handling of light and space can be most impressively read in the house he designed here.

I visited the Esherick House for the first time in the summer of 1980. Even then I was sure that I would return to this house to learn and experience more about it. This home appeared full of secrets that I wanted to discover. Now, in retrospect, I know why this creation of Kahn's occupied me to such a degree: because of changing light and the progression of light throughout the year, it is a house which reveals a different side of its character upon every visit.

分析草图
乌尔斯·布提克 1983

**Analytical sketches
U.B. 1983**

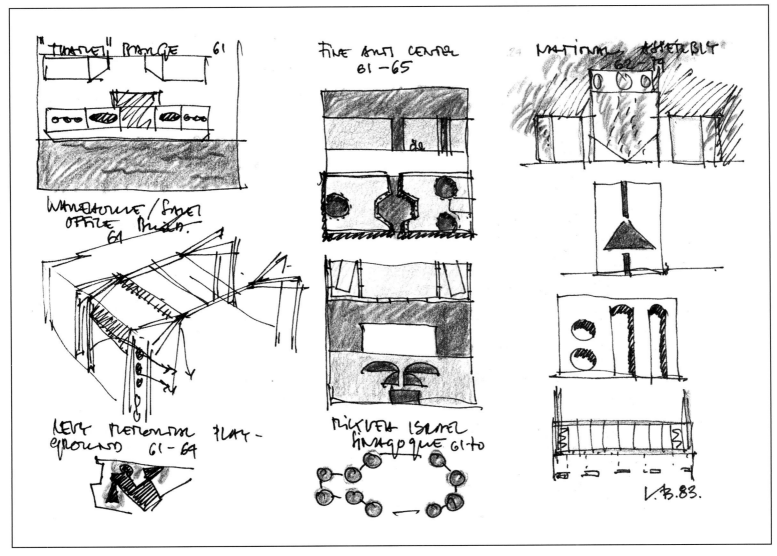

I decided to return to Philadelphia later and to use the Kahn Archives at the University of Pennsylvania to study the oeuvre of Louis I. Kahn. I wanted the time to become closer acquainted with his work through original drawings, photographs, correspondence, and information from earlier collaborators of his.

My original plan to compile comprehensive documentation only on the Esherick House, important as it was in Kahn's work, had to be abandoned quite soon, for good reason. After looking through the original sketches and microfiche evidence under the key word "Esherick House," I discovered to my great surprise that there was, initially, a planned addition to the existing Esherick House and, secondly, an atelier actually constructed for the artist Wharton Esherick. How was it possible that these two projects had remained undiscovered, until almost ten years after Kahn's death, although a book with the title *Louis I. Kahn: Complete Work 1935 – 74* had already appeared?

This discovery, more coincidental than anything else, prompted me to systematically examine all the 25,000 microfiche documents on Kahn's work. While I sat in front of the screen of the "Read and Copy Machine," I discovered a goodly number of Kahn projects which had not been previously published. By photocopying the microfiche documents, I assembled an almost complete archive of Kahn's planned and actually executed buildings. But how was I to analyze and further employ such a mass of new findings?

Finally, I decided to broaden my original idea, i.e., of investigating the Esherick House with respect to light control, design of the windows, the adjustable views from the inside outward, the integrated ventilation panels, the sunshades, and its other components. Instead, I expanded the scope of my project to cover a number of selected designs, and I soon arrived at a definition of my research topic: "Light and space in the works of Louis I. Kahn." I then felt that this clearly defined context would enable me to systematically investigate and to compare Kahn's projects.

Well-briefed by Kahn's earlier colleagues and equipped with a good map of the city, I struck out to find and examine buildings of Kahn's which had until then remained unknown. I took pictures of them and learned as much as I could about them.

During 1982 and 1983, I put together 24 files which, together with help from analytical drawings, clarified the central topic of light in the architecture of Louis I. Kahn. My photocopies of the microfiche documents served as basis for this personal study. Use of overlay techniques revealed those architectural elements which involve the topic of light. The summary texts on each of the

分析草图
乌尔斯·布提克 1983

**Analytical sketches
U.B. 1983**

buildings investigated in this manner is my abstract, individual representation and interpretation of the insights which I gained.

In addition to drawings and collages, my work also resulted in production of a videotape with the title "One Day in the Life of a Window." With the aid of video techniques, I made exposures at the west livingroom window in the Esherick House at intervals of 30 minutes, in the attempt to record the play of light and shadows made possible by this window throughout 24 consecutive hours.

In 1985, my third Kahn-trip took me to India, Nepal, and Bangladesh. Inspecting and photographing the National Capital of Bangladesh proved to be the most difficult task of all my trips. I spent many days of my stay waiting in government offices for the necessary officials, and in explaining to them over tea and cake what I had in mind. After two weeks I finally found the right official, and I received permission to photograph in relative peace a number of the rooms of the Capitol normally off-limits to outsiders.

The Dhaka project was Kahn's most exciting and most complicated work which makes it all the more unfortunate that views of the individual, magnificent rooms of this fantastic complex have until now not been cleared for publication.

Until the moment when I was allowed for the first time to enter the Assembly Hall, I didn't even know how Kahn had solved the problem of light control above the parliament chamber (it was not possible to obtain permission to examine the construction drawings in 1983). I was therefore all the more astonished and enthused when I finally stood in this gigantic chamber, thirty meters high, and beheld how sunlight was reflected in and modulated at the same time from the exterior ring, through semicircular openings, and into the Assembly Hall.

Final material was collected for this book during my last trips to Tel Aviv in winter of 1991, and to Philadelphia in autumn of 1992.

The fifty projects I have selected from Louis I. Kahn's oeuvre for publication in this book demonstrate the extensive variety of techniques which he investigated and employed for the control and modulation of natural light. For each particular building, he attempted to find the individual coherent solution to the problems with sunlight which arose. Kahn considered the window to be an instrument amenable to fine tuning, one which could be designed to effectively provide the interface between outside and inside in ever new and changing variations.

分析草图
乌尔斯·布提克 1983

**Analytical sketches
U.B. 1983**

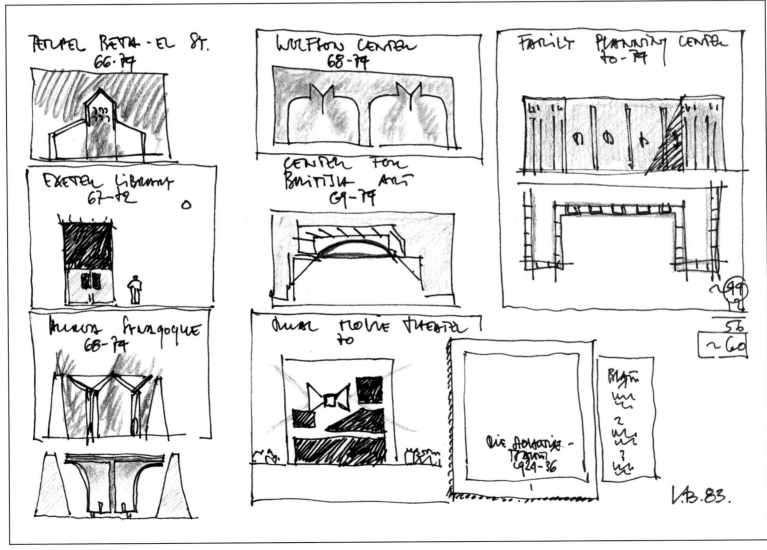

"A space can never reach its place in architecture without natural light. Artificial light is the light of night expressed in positioned chandeliers not to be compared with the unpredictable play of natural light...." (10)

The many and various methods of light control, and the different elements for light modulation, can be traced as well as mutually compared as a leitmotiv passing through Kahn's design work from beginning to end. In my analysis, I have progressed chronologically through the projects, beginning with one of Kahn's student projects from 1924, and ending with the Dhaka capitol project in 1974. My own intention throughout was to reveal and to record Kahn's untiring searches for new solutions in handling the fall of light into a room.

The window from two old words for "wind" and "eye" in its great variety of design forms represents the focal point of my study: the window as wall or ceiling opening through which light (and air) can enter the room. Louis I. Kahn spent his entire professional life in involvement with this element of illumination and ventilation. He considered the individual window to be a formal and functional entity, one which, in turn, became a highly essential component of his spatial structure.

"A plan of a building should be read like a harmony of spaces in light. Even a space intended to be dark should have just enough light from some mysterious opening to tell us how dark it really is. Each space must be defined by its structure and the character of its natural light." (11)

In his early work, Kahn resorted to solutions, more or less conventional, which took the form of separate, protruding sunshade elements. In his later projects, however, he invented and developed a number of integral sunshading measures which have endowed his architecture with its unique and intensely personal character. Kahn's many different experiments also included the functional and formal integration or separation of window and ventilation openings.

Archaic openings and forms particularly characterize the architectural impression made by those of Kahn's buildings which were subjected to extreme climatic conditions, and for which sunlight must be filtered such as his work for Angola, India, Pakistan, Bangladesh, and Israel. Giant "light filters" were designed here which on the one hand furnish shadow and, on the other, ensure sufficient ventilation and a view toward the outside.

"The order of light was respected in the buildings I did in India and Pakistan, now Bangladesh, by giving the outside building to the sun and the interior building to habitation." (12)

分析草图
乌尔斯·布提克 1983

**Analytical sketches
U.B. 1983**

In the late work of Louis I. Kahn, it becomes apparent that skylight plays an increasingly important role, and that Kahn develops ever more sophisticated solutions for light control and modulation. The culmination of all Kahn's experiments is without doubt the Assembly Hall in Dhaka. Here, we quite vividly see how Kahn incorporated the experience gained from his earlier designing and building, in the creation of a genuine architectural jewel.

This book was organized in accordance with the sequence of my own steps of work. The sketches at its beginning (and those for the Esherick House) are intended to show nothing other than an approach, through drawing, to those buildings which I analyzed, a process of approach from which my theme of light originated and gradually developed.

The layout of this book as published reflects the content of the 24 files which were assembled in 1982 and 1983. The documents (plans and sketches) are presented on the left page of the book, and my personal interpretations are found on the opposite page: at the top right, a small emblem as abstraction of the theme "light and space." Page 39 explains the analytical technique with original plan, overlay plan, photograph, as well as emblem with the symbols employed.

For the projects which followed the Esherick House, more space was allotted to the photographs, and detailed drawing analyses were no longer provided.

The end of the book contains a typology of the various light control and modulation elements employed by Kahn.

"Material lives by light," wrote Louis I. Kahn. "You're spent light, the mountains are spent light, the trees are spent light, the atmosphere is spent light. All material is spent light." (13)

分析方法
a. 路易斯·I.康的原始图纸
b. 分析草图
c. 照片
d. 图示
e. 符号

Analytical technique
a. Original plan by Louis I. Kahn
b. Analytical drawing
c. Photo
d. Emblem
e. Symbols

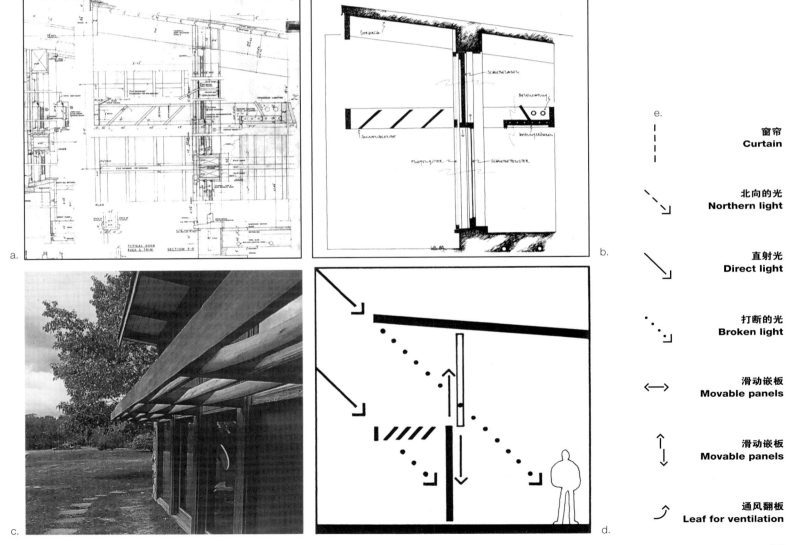

学生时代作品
购物中心
1924

STUDENT WORK
SHOPPING CENTER
1924

据我所知，这是仅知的路易斯·I. 康在保罗·克瑞教授指导下完成的两个学生作业之一。这个设计完全符合鲍扎体系理念。正立面设计依据传统的三段式原则，分为台基、主体和屋顶部分。康的设计严格遵守轴对称结构，没有特定的地理朝向。在首层，康在四面均设计了门廊。七层通高的内庭走廊最初可能想要覆以玻璃屋顶。走廊将建筑分为两个商铺区和两个办公区，区域之间在两个主入口上部以廊桥相连。容纳了楼梯、电梯和辅助用房的服务区位于两个矩形建筑单元之间。拱形走廊通往办公室。两个办公区的窗户似乎想要设计成从各楼层一直连通至屋顶。为了遮蔽直射阳光而设置的遮阳构件并不明显，这突出了垂直壁柱的纪念性。仅在廊桥区域，能够利用双层墙体对光线进行调节（参阅右页图）。

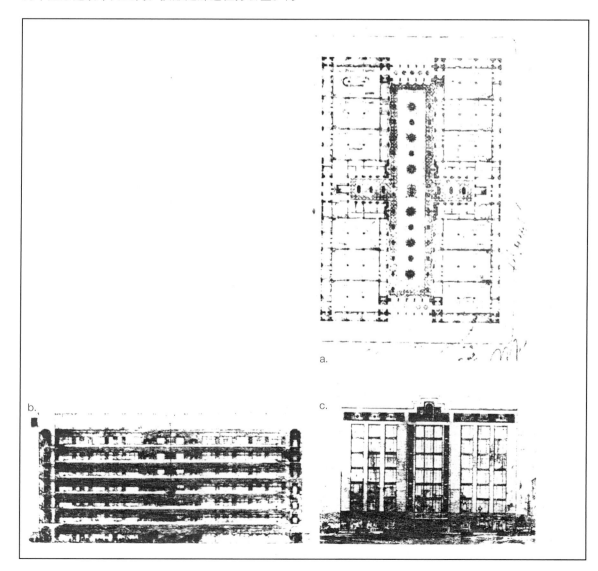

a.

b.

c.

To the best of my knowledge, this is one of the only two projects known from Louis I. Kahn's student work under Professor Paul P. Cret. It is entirely in the spirit of the Ecole des Beaux-Arts. The facade is designed according to the three-fold classical breakdown into base, main order, and attic. Kahn's project features austere axial-symmetrical configuration, with no particular orientation to geographical directions. On the ground floor, Kahn integrates a portico which appears in all four facades. An interior-court gallery extends through all seven floors and is perhaps intended to be covered by a barrel-vaulted glass roof. The gallery divides the building into two shop and office blocks, which are bridged over the two main entrances. The stairways, elevators, and auxiliary rooms are configured as servicing elements in the middle of the two oblong building units. Arcade corridors provide access to office rooms. It appears that the windows in the two office blocks are intended to extend from the floors to the ceilings. Shading elements for protection against direct sunlight are not apparent, a lack which allows the pilasters prominence in their monumental verticality. It is only in the area of the access bridges that modulation of light is possible, by virtue of the double walls (see drawing on the right page).

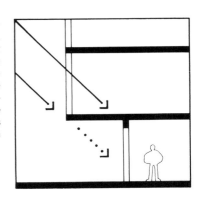

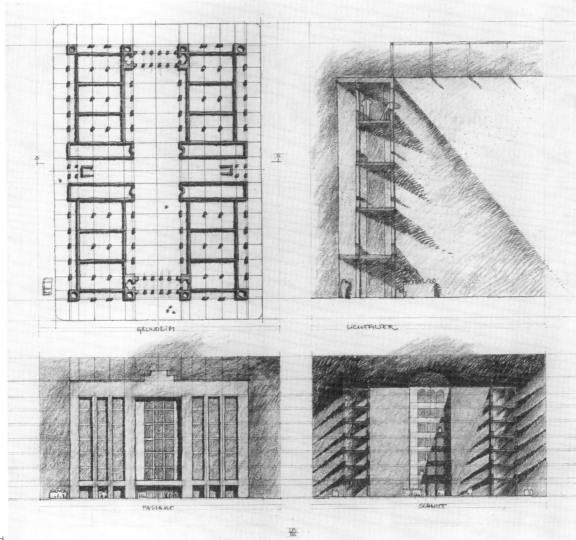

a. 平面图
b. 剖面图
c. 立面图
d. 分析草图

a. Plan
b. Section
c. Elevation
d. Analytical drawing

泽西住宅项目
1935–1937；建成
罗斯福自治市，
新泽西州
与阿尔弗雷德·卡斯特纳合作
完成

JERSEY HOMESTEADS
1935 – 1937; BUILT
ROOSEVELT BOROUGH,
NEW JERSEY
WITH ALFRED KASTNER

泽西住宅项目为许多从欧洲移民而来的犹太家庭提供了住所。这个社区中主要有大约200个住宅、一所学校（同时作为社区中心）、一个现代服装厂以及一个农场。

这些度假屋式的平房建造在公园般的景观中间，与一条环形的乡村道路相连，在当时也算是非常宽敞的。大部分住宅是单层的，没有地下室。墙体由混凝土砌块砌成，非独立的平屋顶是一片略微出挑的混凝土板。从其建造技术和建筑语言可以看出，这是康第一次（在真实环境中）尝试在大尺度建筑上采用现代建筑原则，检验其影响，并在应用中总结经验。每间住宅的外形是内部房间结构的体现，内部房间被清晰地分为日间与夜间区域。长长的车库附加在楼层平面的一侧。餐厅和起居室的落地窗一直延伸到天花板。窗户有些部分是不能开启的，有些挂有扉帘。康没有在此处设置遮阳板遮挡照射在大块玻璃上的直射阳光。其他窗户就像是印在墙上的。窗洞口上方有一道小过梁。

b1.

b2.

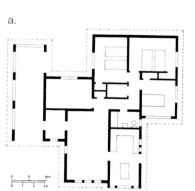
a.

The development Jersey Homesteads became the new home for many Jewish families who emigrated from Europe. This cooperative essentially consisted of about 200 homes, a school (also serving as community center), a modern plant for production of clothing, and a farm.

The bungalow-type homes were built in park-like landscape, adjacent to a circular village street, and were quite spacious for their time. Most were one-story, and none had basements. The walls consisted of concrete blocks, and the non-isolated flat roof is a slightly protruding concrete slab. The construction technology as well as the architectural language represents Kahn's first attempt (in an actual context) to implement the principles of modern architecture on a large scale, to test their effects, and to gain experience in their use. The outward form of each home is the result of the configuration of rooms inside, which have been clearly broken down into daytime and nighttime areas. The long garage appears to have been added on at the side of the floor plan. The windows of the dining and living room extend from the floor to the ceiling. Some window parts are non-opening, and some have casements. Kahn installed no sunshades here to protect the large glass surfaces from direct sun. The other windows appear to have been stamped out of the walls. A small lintel spans the window opening.

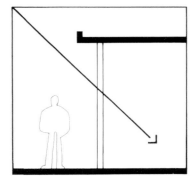

a. 平面图
b1./b2. 居住单元
c. 窗户细部

a. Plan
b1./b2. Housing units
c. Window detail

艾哈瓦以色列犹太教会堂
1935—1937；建成
费城，
宾夕法尼亚州

AHAVATH ISRAEL
SYNAGOGUE
1935 – 1937; BUILT
PHILADELPHIA,
PENNSYLVANIA

路易斯·I. 康第一个独立设计的作品建成时，他已经34岁了。（当时他同时与阿尔弗雷德·卡斯特纳合作，在新泽西州罗斯福自治市进行泽西住宅项目设计。）这个早期设计的平面图，试图在犹太教会堂中应用现代主义设计原则，但图纸已不复存在了。这座犹太教会堂是一个纯净的平顶立方体，立于一排排的传统建筑之间。砖制的入口立面设计成一面巨大的墙。有三个小窗户为楼梯间提供采光。南向墙面也是砖墙，是完全封闭的。然而北立面涂了灰泥，可见的承重结构和非承重的填充墙，令人想起工业建筑结构。北立面上的窗户为门厅、首层的犹太教会堂以及楼上的画廊提供采光。窗户一直延伸到屋顶，构成"屋顶框定"的窗，这是一种经济的解决方式，因为屋顶也同时作为窗的过梁。室内出人意料地温暖，氛围也截然不同：在室内，北立面工厂般的特质似乎被忽略掉了。木质墙板嵌入结构布局中，精确地留出筒灯和通风口，与北向的窗户交相呼应，使房间沐浴在温暖的光线之中。房间靠庭院的墙面一侧，是讲经台的位置，此处采用了玻璃砖。

a. 入口立面
b. 内部

a. Entrance facade
b. Interior

a.

It was as a 34-year-old architect, that Louis I. Kahn was first able to build on his own. (At the same time he also collaborated with Alfred Kastner on the Jersey Homesteads in Roosevelt, New Jersey.) The plans for this early design, an attempt to implement concepts of modernism in a synagogue, are no longer extant. This synagogue, a pure cuboid with a flat roof, stands in a conventional line of row buildings. The brick entrance facade is designed as a massive face wall. Three small windows provide light for the stairwell. The south longitudinal facade, also brick, is completely closed. The north facade, however, is plastered and calls an industrial structure to mind, with its visible supports and non-loadbearing infilling walls. The north facade contains windows which furnish light for the foyer, the synagogue chamber on the first upper floor, and the gallery above. The windows extend to the ceiling, creating "ceiling-framed" windows, an economical solution since the ceiling also provides the lintel. The interior radiates surprising warmth and a distinguished atmosphere: inside, the industrial character of the north facade appears forgotten. Wood paneling integrated into the structural configuration features precisely placed spotlights and ventilation openings in interplay with the massive north windows, and bathes the room in warm light. The wall on the courtyard side of the room, where the Bima (lectern) is located, is provided with glass brick.

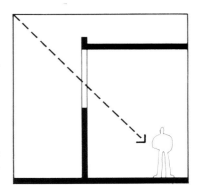

b.

预制住宅研究
1937—1938；未建成
与路易斯·马加齐纳及亨利·克隆布共同设计

PREFABRICATED HOUSE
STUDIES
1937 – 1938; UNBUILT
WITH LOUIS MAGAZINER
AND HENRY KLUMB

这个研究项目受到了塞缪尔·锡米恩·费尔斯（Samuel Simeon Fels）的支持，费尔斯曾表达出他对经济型建造模式的兴趣。当时，距离路易斯·I. 康获得学位已过去十三年。为何其间有如此长时间的中断？一方面，康在相当长的一段时间内在不同的建筑师事务所中工作；另一方面，这个项目是在大萧条时期，当时的建造项目相当低迷。

康的"预制住宅"，是他为自己和几位同事布置的任务，是他自己对于社会问题思考的结果。此处我们看到的是将现代建筑的整体目标在大批量生产的住宅中进行的初次尝试。典型的联排住宅有平屋顶和特大的窗户，来满足人们对阳光、空间和空气的期许。在底层，窗户占比很大，而且充分暴露在阳光（和严寒）中，或许这就是为何平面图中特意画出了窗帘。住宅中有传统的门廊（但没有任何装饰）。传统的双悬窗第一次被竖向窗挺的带形窗取代。室内的浴室通过一扇形式传统但超大尺寸的天窗采光。我的图示中想要表现出在整个墙面上不成比例的窗户洞口。

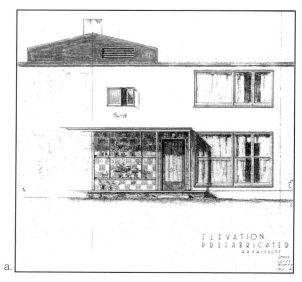
a.

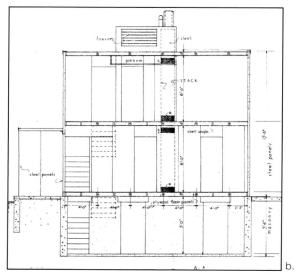
b.

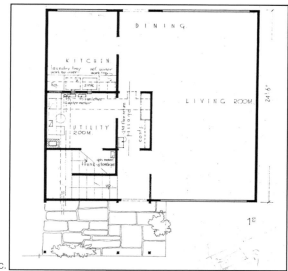
c.

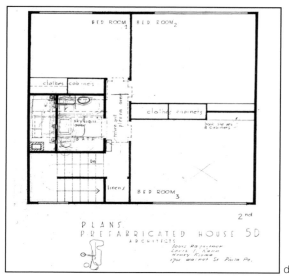
d.

This study project was supported by S. Fels, who had voiced his interest in an economical mode of construction. By now, thirteen years had passed since Louis I. Kahn had received his diploma. Why this long interruption? For one reason, Kahn had worked for a relatively long time in various architects' offices; for another, this was the Great Depression, during which construction was at low ebb.

Kahn's "Prefabricated House", an assignment which he posed to himself and a number of his colleagues, was the result of his own thinking on social issues. Here we see the first attempts to integrate objectives of modern architecture into mass-produced housing. The typical row house is provided with a flat roof and extra-large windows, in order to meet expectations for light, space, and air. On the ground floor, the windows are over-proportioned and are fully exposed to light (and cold), perhaps the reason why curtains are shown in the plans. The traditional porch is employed (but without any embellishment). For the first time, the otherwise customary double-hung (sash) windows were replaced by strip windows with vertical muntins. The bathroom, in the interior, receives light through a conventional but overdimensioned skylight. My schematic drawing attempts to show the disproportional ratio of the window openings to the total wall area.

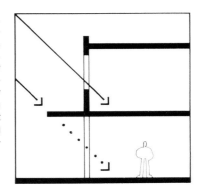

a. 立面图
b. 剖面图
c. 底层平面图
d. 顶层平面图
e. 分析草图

a. Elevation
b. Section
c. Ground floor
d. Top floor
e. Analytical drawing

杰西·奥泽住宅
1940—1942；建成
埃尔金斯·帕克，
宾夕法尼亚州

JESSE OSER HOUSE
1940 – 1942; BUILT
ELKINS PARK,
PENNSYLVANIA

康的第一个独栋住宅以多样的建筑变化为特征。在艾哈瓦以色列犹太教会堂中呈现的结构和形式单元在此处被尝试性地抛弃掉了。康在这个平屋顶住宅中将原石砌体和木结构结合在一起，为此他采用了大量不同的窗户类型和形式。东北立面最有特征的是一个挑出于住宅基座上方的木质盒子，上面有两个上下排列跨过转角的带形窗。此处，康首次否定了"旧的"荷载与支撑的法则，通过"新的"现代主义建筑的典型方式来表达轻盈与失重。水平的窗户在整个立面部分延伸，更强调了这个意图（图中样式 C：参阅右页分析草图）。在底层（西南面），巨大的角窗像温室玻璃一样环抱着起居室。内部与外部之间"流动空间"的概念第一次在此处呈现（图中样式 D）。嵌入原石砌筑墙体的窗户按照内部房间的功能布置，相互之间没有关联。在底层，它们就像是从砌体中打出的洞口一样（样式 B）。在上层，这些窗户延伸到屋顶（样式 A："屋顶框定"的窗户）。示意图表现了转角窗（样式 D）。

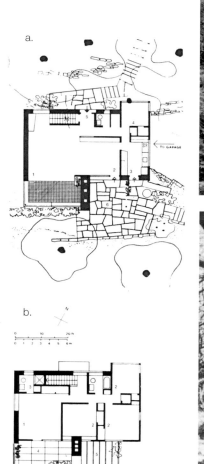

c1.

c2.

Kahn's first single-family house is characterized by exceptional architectural variety. Structural and formal unity as in Ahavath Israel Synagogue appears discarded here in favor of experiment. Kahn combined rough-stone masonry work with timber construction for this flat-roofed house, for which he employed a great variety of window types and forms. The northeast facade is characterized by a timber-constructed cube protruding beyond the house footing, and by two window strips, one above the other, which wrap around the corners. Here, for the first time, Kahn negates the "old" rule of loads and supports in order to evoke effects of lightness and weightlessness through "new" approaches characteristic of Modernism. The horizontal window extending over the entire facade section reinforces this intention (Type C: see analytical drawings to the right). On the ground floor (southwest), the huge corner window wraps around the living room like conservatory glazing. The concept of "flowing space" between inside and out appears here for the first time (Type D). The windows, recessed into the rough-stone masonry wall, are located and dimensioned according to the functional conditions of the interior rooms, and are not interrelated. On the ground floor, they give the appearance of being punched out of the masonry (Type B). On the upper floor, they extend to the ceiling (Type A: "ceiling-framed" windows). The emblem shows the corner window (Type D).

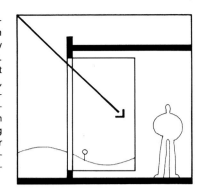

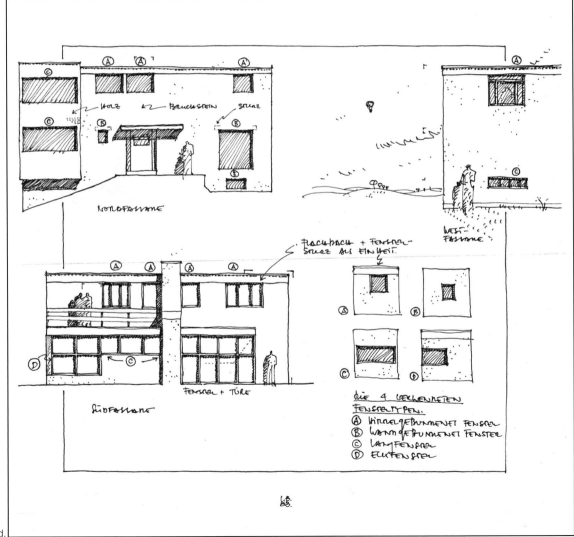

a. 底层平面图
b. 顶层平面图
c1./c2. 外部
d. 分析草图

a. Ground floor
b. Top floor
c1./c2. Exteriors
d. Analytical drawing

L. 布鲁多住宅
1941—1942；未建成
埃尔金斯·帕克，
宾夕法尼亚州

L. BROUDO HOUSE
1941 – 1942; UNBUILT
ELKINS PARK,
PENNSYLVANIA

原计划布鲁多住宅紧邻奥泽住宅建造。这个住宅是在奥泽住宅一年后设计的，相较于奥泽住宅，它看起来更传统。它是一座木质住宅，有水平的墙板、人字坡屋顶，前面设有一个车库，与住宅之间形成一定夹角（是由于基地的几何形状以及建筑朝向正南造成的）。然而进一步观察，就能发现一个康将应用多年的构件：遮阳板，此处是水平遮阳板。装有带形窗的南立面（窗户与砌体的面积比例大约为2:1）上就装有一个遮阳板。此处窗户也一直延伸到了屋顶。遮阳板只是一个由悬臂支架和倾斜的企口板组成的简单木质构件。在厨房前面，餐厅和起居空间的遮阳板延伸成藤架顶棚的样子（参阅上层平面图），为外面的座椅提供遮阳。为了给上层的窗户遮阳（两个正方形窗和一个带形窗），非对称的坡屋顶偏向南侧。此处的出檐似乎有点太小，无法满足遮阳功能，尤其是对南立面左侧的大方窗来说。在后面的设计中，康会继续致力于带形窗上方屋顶出檐的尝试。

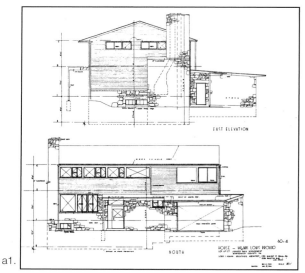
a1.

a2.

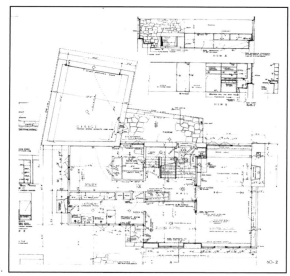
b.

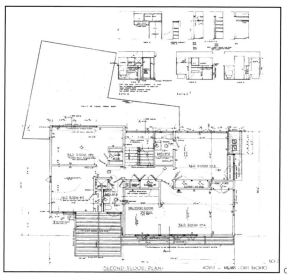
c.

The Broudo House was intended to be built very near the Oser House. The Broudo House, designed a year later, appeared conventional in comparison. A timber house with horizontal siding, gable roof, and a garage situated at the front at somewhat of an angle (owing to the geometry of the property and the orientation of the house to the south). Upon closer examination, however, one finds an element with which Louis I. Kahn will be involved for years: the brisesoleil, configured here horizontally. The south facade, provided with a strip window (ratio of window area to masonry area approx. 2 : 1), is protected by one of these elements. Here as well, the windows extend all the way to the ceiling. The brisesoleil is a simple timber structure with cantilevered brackets and rebated boards set at angles. In front of the kitchen, the brisesoleil of the dining and living space is extended to form a type of pergola roof (see plan of upper floor), and provides shade for an outside seating area. In order to shade the window areas on the upper floor (two windows in square form, and one strip window), the gable roof projects asymmetrically toward the south. The projecting roof covering here appears too small to provide sunshade function, especially for the large square window at the left in the south facade. Kahn will continue to involve himself in ongoing experiments with the topic of a roof projecting over a strip window.

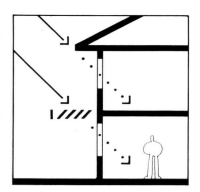

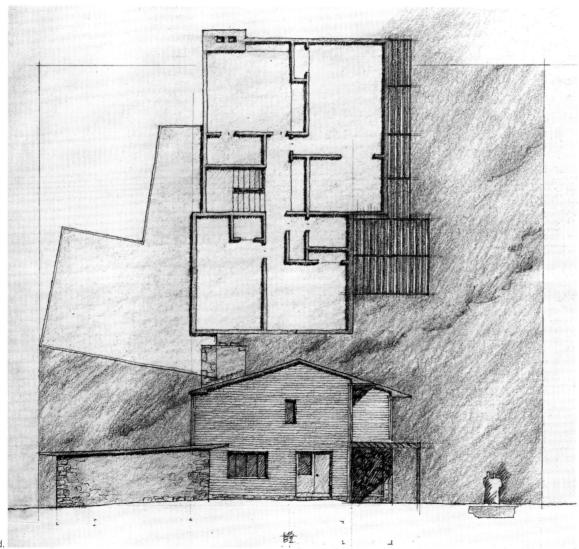

a1./a2. 立面图
b. 底层平面图
c. 顶层平面图
d. 分析草图

a1./a2. Elevations
b. Ground floor
c. Top floor
d. Analytical drawing

卡佛庭院住宅开发项目
1941—1943；建成
卡伦镇，
宾夕法尼亚州
与乔治·豪及
奥斯卡·格雷戈里·斯托罗诺
夫合作完成

CARVER COURT
HOUSING
DEVELOPMENT
1941 – 1943; BUILT
CALN TOWNSHIP,
PENNSYLVANIA
WITH GEORGE HOWE AND
OSCAR GREGORY STONOROV

这个开发项目位于一个伤残军人医疗中心附近，起初是为医疗中心员工提供住宿的。卡佛庭院只是康在第二次世界大战期间快速建成的几个项目之一。因此康有机会继续探索他自己的建筑语言。在所有建成与未建成的战争住宅项目（松福德庄园、斯坦顿道、百合池塘、彭尼帕克森林、浆果庄园、柳木飞机制造厂、轰炸机城市以及林肯公路国防住房）中，卡佛庭院是最有趣的一个。

此处只有两种户型。独立的四户住宅因其光秃秃的立方体外观显得格外有趣。主立面仅在底层被一条带形窗打断（依据原始平面图），然而这条带形窗最终换成了两个小方窗。

联排住宅首层采用过大的砌体横墙承重，仅容纳了车库、入口区域及锅炉房。起居室和卧室在楼上。底部的垂直性与安装在水平墙板上的带形窗形成对比。带形窗由一系列并列安装的传统双悬窗组成。屋顶的出挑没有明确的形式或功能目的：作为遮阳板不够有效；作为装饰线脚，屋顶出挑的效果又太弱了。

a.

b.

This development, located near a medical center for wounded veterans, originally served to accommodate the medical staff. Carver Court is only one of several of Kahn's developments very quickly executed during World War II. Kahn therefore had the opportunity to continue searching for his own architectural language. Of all the merely planned or actually executed war housing projects (Pine Ford Acres, Stanton Road, Lily Ponds, Pennypack Woods, Berry Farms, Willow Run, Bomber City, and Lincoln Highway Defense Housing) Carver Court is one of the most interesting.

Only two house types can be treated here. The freestanding four-family house is intriguing due to its stark, cubic appearance. The main facade is interrupted on the ground floor only by one strip window (cf. original plan) which, however, was eventually replaced by two small square windows.

The row house stands on overly massive masonry cross walls, which accommodate only garage, entrance area, and furnace room. The living and bedroom areas lie above. The verticality of the footing section contrasts with the window strip installed in the horizontal clapboard siding. The window strip consists of a series of conventional double-hung (sash) windows installed next to each other. The slightly protruding roof serves no convincing formal or functional purpose: it is ineffective as brisesoleil and, as fascia, the extending roof makes too weak an effect.

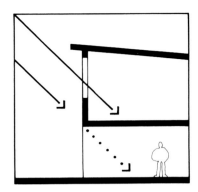

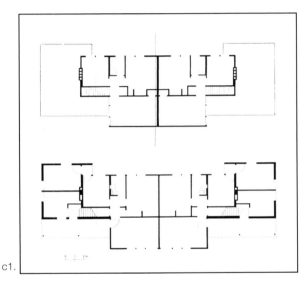

c1.

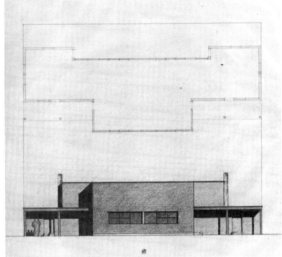

d1.

c2.

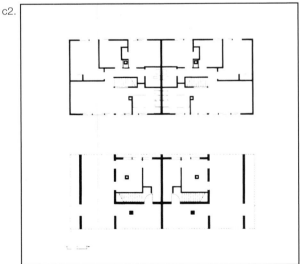

d2.

a. 南立面（户型 A）
b. 南立面（户型 C）
c1./c2. 平面图（底层及顶层平面）
d1./d2. 分析草图

a. South elevation (Type A)
b. South elevation (Type C)
c1./c2. Plans (ground and top floors)
d1./d2. Analytical drawings

194X 旅馆
1942—1943；未建成
与奥斯卡·格雷戈里·斯托罗诺夫合作完成

HOTEL FOR 194X
1942 – 1943; UNBUILT
WITH OSCAR GREGORY STONOROV

1942 年，康和斯托罗诺夫受到《建筑创作》杂志的邀请，完成一项关于预制结构的设计。尽管两位建筑师原本没有赶上最后的出版期限，一年后，他们还是找到机会完成了这个项目，并投稿给了同一家杂志。他们完成了一个 20 世纪 40 年代典型的旅馆设计：因此项目名称叫做"194X 旅馆"。

公共空间在首层，受到上部十三层体块的支配。标准层平面按照传统方式设计：客房沿走廊东西向排列。柱子支撑着的混凝土楼板靠近边缘处逐渐变薄。于是，每个房间的空间向外逐渐变得更高也更明亮。

除此之外，这个构造可以允许采用横向通长的带形窗。遮阳板装在窗前，用来削弱直射阳光的作用。这个外形坚固的水平构件，是由铝制波纹板构成的，板上打上了矩形孔洞。这些孔洞随着太阳高度的变化在立面上创造了一种几何形的光影变幻，这个设计与五年后建成的拉德维尔治疗大楼有相似的效果。

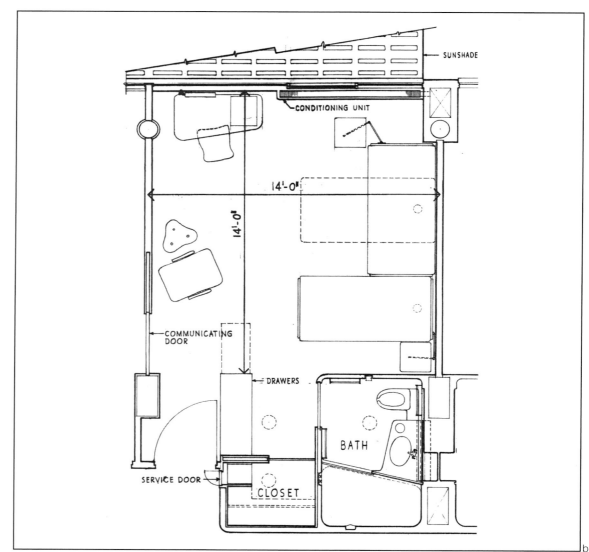

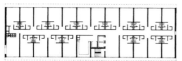

a.

b.

In 1942, Kahn and Stonorov were invited by the journal *Architectural Forum* to submit a design dealing with the topic of prefabrication. Although the two architects were not originally able to meet the deadline for publication, they found the opportunity one year later to complete the project and submit it to the same magazine. They submitted a design for a typical hotel of the 1940's: hence the name "Hotel for 194X."

The public spaces are accommodated in the ground floor, which is dominated by the thirteen-story block installed above. The floor plan of the floors is conventionally designed: the guest rooms are lined up in an east-west direction along the corridor. The concrete floor slabs, supported by columns, are designed as brackets tapering toward the outside. As a result, the space in each room becomes higher and brighter toward the facade. In addition, this construction allows use of strip windows, which extend over the entire length of the cube. Brisesoleils are installed in front of the windows to diminish the effects of direct sunlight. This formally strong, horizontal element consists of corrugated aluminum sheeting perforated by small rectangular slots. These cut-out openings create a geometric play of lights and shadows on the facade, according to the height of the sun, which point to similar effects in the Radbill Therapy Building built five years later.

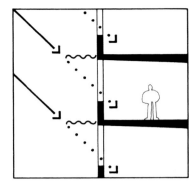

a. 标准层平面图
b. 房间平面图
c. 分析草图

a. Typical plan
b. Plan of a room
c. Analytical drawing

国际女装工人工会健康中心
1943—1945；未建成
费城，
宾夕法尼亚州
与奥斯卡·格雷戈里·斯托罗诺夫合作完成

INTERNATIONAL LADIES
GARMENT WORKERS
UNION HEALTH CENTER
1943 – 1945; UNBUILT
PHILADELPHIA,
PENNSYLVANIA
WITH OSCAR GREGORY
STONOROV

这个项目的任务是重新设计工会大楼的入口，这是康第一次面对一个现存建筑，而且是一座按照古典原则设计的砖造建筑。康计划完全重新设计这幢古典建筑对称的入口。他在一面大橱窗上安装了巨大的钢过梁，门被移动到偏离对称轴的位置。门窗四周的框采用了白色大理石，像一个盒子一样从立面上悬挑出来。（楼梯和平行于建筑正面的休息平台的饰面，在康后来设计耶鲁美术馆时再次应用在入口设计中。）尽管首层窗户的调整没有改变结构体系，但窗户的垂直布置方式由水平布置所取代。另一个大理石框也从立面上挑出来，突出了新的"假"带形窗。基于完全不同的材料感受，康有意试着将传统砖造建筑与现代建筑理念并置在一起。不同建筑理念与建筑语言之间的冲突在此处更加明显了。

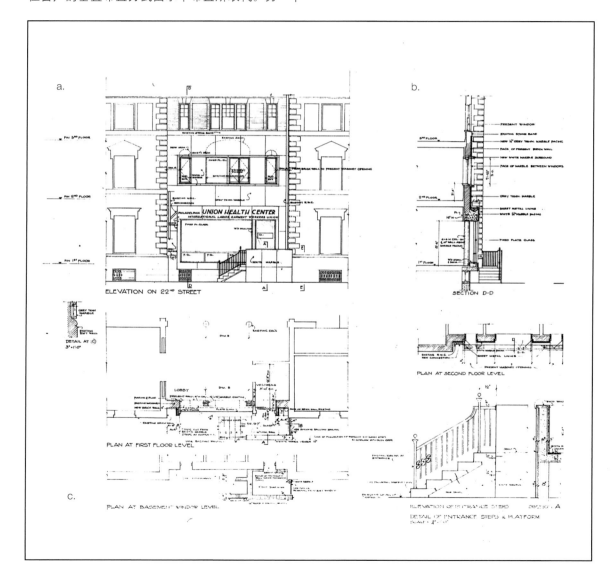

a. 立面细部
b. 剖面图
c. 平面图
d. 分析草图
e. 西立面

a. Detail of elevation
b. Section
c. plan
d. Analytical drawing
e. West facade

The task here was to redesign the entrance to the Union building, the first time that Kahn was confronted with an existing building and, at that, a brick building designed according to classical rules. Kahn planned to completely redesign the symmetrical entrance of the classicistic building. He employs a huge steel lintel in installation of a large show window, with a door shifted out of the symmetry axis. A white marble frame surrounds the window and door, and projects boxlike from the facade. (The stairs and facing of the landing parallel to the building frontage line anticipate the entrance Kahn later designed for the Yale Art Gallery.) Although the modifications to the windows on the first upper floor result in no structural-engineering changes, the vertical window configuration gives way to the horizontal. Another marble frame, also protruding from the facade, emphasizes the singularity of the new "false" strip window. Kahn consciously attempts to confront the traditional brick building with concepts of modern architecture, based on utterly different materials consciousness. Conflict becomes evident between different concepts and languages of architecture.

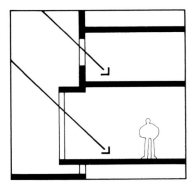

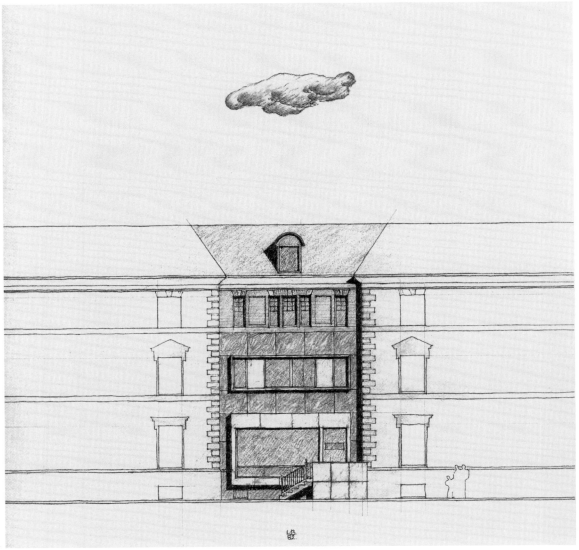

d.

e.

阳伞住宅
1944；未建成
与奥斯卡·格雷戈里·斯托诺罗夫合作完成

PARASOL HOUSES
1944; UNBUILT
WITH OSCAR GREGORY STONOROV

在康（20世纪）40年代早期的各种设计中能够看出，他在尝试将现代主义建筑中的标志性元素吸纳进自己的建筑语言当中。这一次，他的设计仍是无法实现的虚拟作业，这也使得他可以去尝试新的建筑理念。似乎此时鲍扎体系的理念已经远远被他抛在脑后了。这个设计是基于勒·柯布西耶的设计理念，也令人想起1916年的多米诺住宅和1922年的雪铁龙住宅。非承重的、以奇怪角度布置的墙体，被插入混凝土柱网中。康完全没有柯布西耶将优美的曲线墙与承重柱网形成对比的天分。两层的起居和用餐空间及走廊、以某个角度插入房间中的楼梯、起居室中间奇怪的单独一根柱子，都造成了一种奇怪的效果。两个遮阳构件严重干扰了二层通高窗户的垂直性（原本计划对水平的天花板和屋顶形成对比）。天花板和屋顶边界出挑，在墙面上形成了光影的变化，在设计出挑的混凝土板时就考虑到了这一点。在43岁的年纪，康又开始了一次新的、未知的冒险。

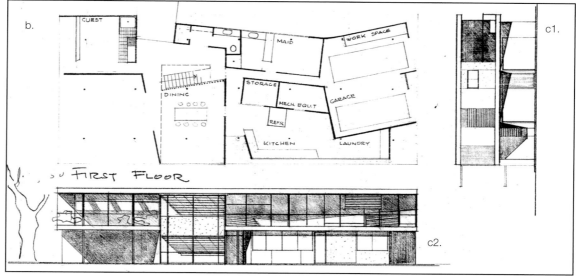

As shown by various of Kahn's designs from the early forties, he attempts here to incorporate programmatic aspects of Modernism into his architectural language. Again, his designs are the result of fictitious assignments free from the expectations of an actual developer which allows him to experiment with new architectural concepts. It appears that ideals of the Beaux- Arts tradition have now been left far behind. The design shown here is based on Le Corbusier's fund of ideas and calls to mind the Domino House of 1916 and the Citrohan House of 1922. Non-loadbearing, strangely angled walls are inserted into a grid of concrete columns. Kahn utterly lacks Le Corbusier's talent for employing elegantly curved walls as contrast to the rigid arrangement of building supports. The two-level living and dining room with gallery, the stairway at an angle into the room, and the one isolated column in the middle of the living room all create an awkward effect. Two brisesoleil elements seriously disturb the intended verticality of the window which extends over two stories (planned to contrast with the horizontal ceiling and the roof).

The setback of the walls behind the ceiling and roof edge creates a play of shadows which takes into account the concept of the projecting concrete slabs. At the age of 43, Kahn stands once again at the beginning of adventure into the new and unknown.

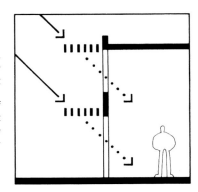

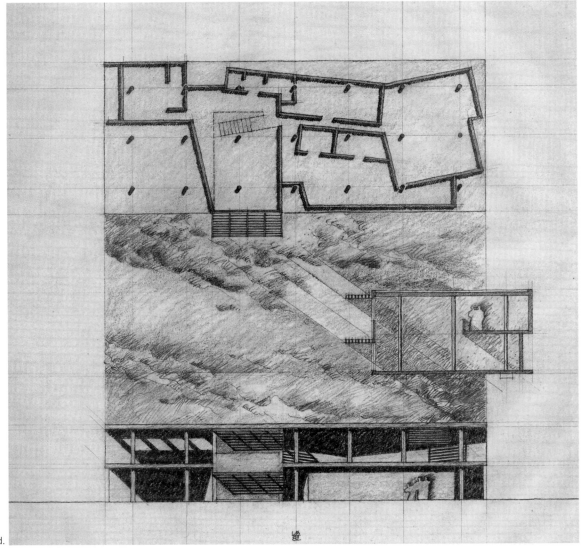

a. 透视图
b. 平面图
c1./c2. 立面图
d. 分析草图

a. Perspective
b. Plan
c1./c2. Elevations
d. Analytical drawing

费城精神病医院
1944—1946；未建成
与奥斯卡·格雷戈里·斯托诺
罗夫合作完成

PHILADELPHIA
PSYCHIATRIC HOSPITAL
1944 – 1946; UNBUILT
WITH OSCAR GREGORY
STONOROV

如果这个扩建项目建成，它将与现有的医院在北侧连接。康规划了一条有顶棚的车道、一个设有接待处的门厅、医生办公室、治疗室、病房以及病人休息室。独特的几何角度是由用地形状决定的。两层高的治疗室位于入口区、研究室和医生办公室的上方。这间治疗室最突出的特征是朝西的大面积的墙面，在人眼高度上没有任何洞口。屋檐下倾斜的窗户投下的光影，在建筑这个部位的立面上形成了独特的边界。显然这个案例中的出挑并非为了遮阳；相反，此处的主要动机是为了形成三段式构图（基座、主体和檐部）。出挑的遮阳构件安装在病房（西北立面）的带形窗上方。相对于窗户的高度来说，遮阳构件似乎实在太窄了。

a. 轴测图
b. 西立面

a. Axonometric projection
b. West facade

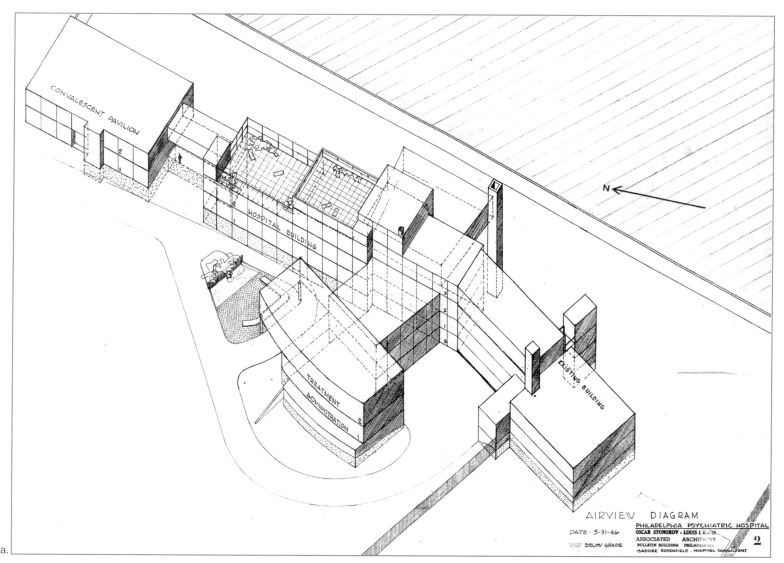

a.

This extension, if executed, would have connected into the north side of the existing hospital. Kahn planned a covered automobile accessway, a foyer with reception, offices for physicians, treatment rooms, ward rooms, and day rooms for the patients. The peculiar angularity to the geometry is the result of the shape of the property. The double-storied therapy room is situated above the entrance area, the staff rooms, and the doctors' offices. A salient feature of this therapy room is the sweeping wall, toward the west, which is without openings up to eye level. The windows installed at a slant under the roof create a shadow which provides a distinct termination to the facade for this section of the building. The projection in this case is surely not intended for purposes of sunshade: instead, the primary motivation here is the concept of threefold subdivision (base, main order, and attic). Projecting sunshade elements are installed in the patients' wards (northwest facade), over the strip windows. The sunshades appear excessively narrow in relation to the height of the window area.

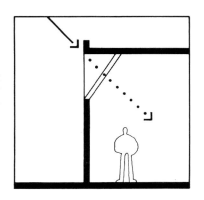

b.

B.A. 伯纳德住宅
1945；加建
金伯顿，
宾夕法尼亚州
与奥斯卡·格雷戈里·斯托诺罗夫合作完成

B.A. BERNARD HOUSE
1945; ADDITION
KIMBERTON,
PENNSYLVANIA
WITH OSCAR GREGORY STONOROV

我认为，伯纳德住宅是路易斯·I. 康设计的最独特的建筑之一。

伯纳德住宅项目是对一幢现存住宅的扩建，住宅原始平面图已经没有了。住宅的基座是由粗糙原石砌筑的，支撑着三个自由边上的木结构。柱子的布置方式尤为奇特：例如，在出挑最大的位置，只有两根柱子；在出挑比较小的位置，却有五根。除此之外，为了承受偏心荷载，在接近顶端的位置，增加了柱子。奇怪的柱子数量肯定让康很不愉快，因为他在图中中间的柱子上打了个叉，写着"否"。建筑上层令人想起卡佛庭院中的联排住宅。南立面上最有特色的地方是通长的横向带形窗。窗挺也很有特点：它们完全不考虑下面的柱子。而且，不同深度房间的窗户都是一样的大小。平屋顶的出挑可以看作是遮阳板。与卡佛庭院不同，墙板是垂直的而非水平的：这种形式当时在美国现代主义者中引发了激烈的争论。随后的改变从根本上改变了这个加建方案。

a.

b.
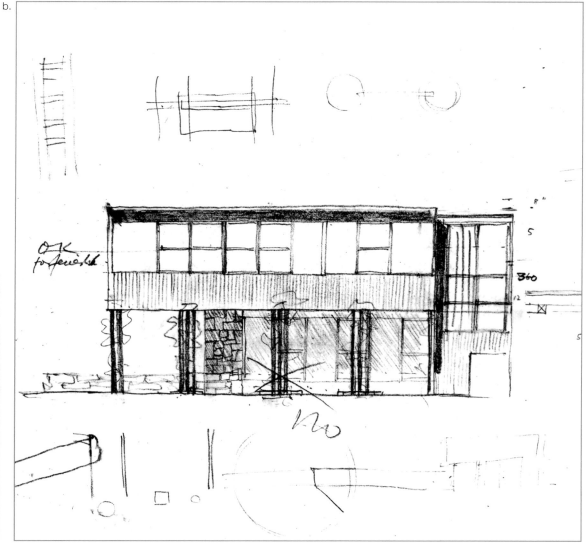

In my opinion, the Bernard House represents one of the most peculiar designs produced by Louis I. Kahn.
The Bernard project entailed an addition to an existing house, the original plans for which were no longer extant. The footing to the house consists of rough stone, which supports a wooden structure protruding beyond the footing on all three free sides. The arrangement of the columns is particularly curious: for example, at positions at which the overhang is greatest, there are only two columns; where the projection is less, there are five. In addition, the columns increase in diameter toward the top, in order to carry the loads by half-over and half-under support. The odd number of columns must have displeased Kahn, since he crossed through the middle one with the comment "No" written on the plans. The upper floor brings to mind the row houses in Carver Court. The distinctive aspect of the south facade is the strip window which extends its entire length. The muntins also represent a peculiar feature: they take no account of the columns below. Moreover, the rooms with different depths all have the same window size. The overhang of the almost-flat roof could be considered a brisesoleil. Unlike the Carver Court project, the clapboard siding is vertically instead of horizontally installed: an aspect which was at that time intensely debated among Modernists in the USA.
Subsequent alterations have radically changed the addition.

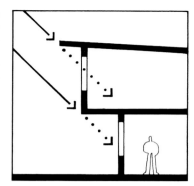

a. 南向视角
b. 南立面图（晚期版本）
c. 分析草图

a. South view
b. South elevation (late version)
c. Analytical drawing

宾夕法尼亚太阳屋
1945—1947；未建成
与奥斯卡·格雷戈里·斯托诺罗夫合作完成

PENNSYLVANIA
SOLAR HOUSE
1945 – 1947; UNBUILT
WITH OSCAR GREGORY
STONOROV

1946 年，利比·欧文斯·福特玻璃公司赞助了一项关于瑟莫潘双层隔热窗玻璃的研究，主要是关于这种玻璃在建筑材料技术、照明工程以及特殊建筑安装技术中的应用。利比玻璃公司邀请了 49 位建筑师参与设计。宾夕法尼亚太阳屋项目采用了新的、面向未来的及已知的构件。最值得注意的特征是北面厚重的砖墙，上面开着小小的浴室的窗子以及入口区域的洞口。东立面和西立面向着太阳方向倾斜；南面也是如此，有大面的连续玻璃墙面。为了给这些大窗户遮挡直射阳光，康设计了四种控制构件：首层上部出挑的遮阳板，安装了倾斜的条板（参阅布鲁多住宅）；平屋顶在顶层的四周远远地伸出（参阅卡佛庭院）以及康新发明的两个构件。一个是西面窗户前的滑动折叠墙板以及（首层及东向和南向立面的）窗户内侧的水平滑动木质嵌板。尽管折叠百叶在康后期的设计中再没有出现过，但是滑动嵌板却以经过调整并改进的形式作为主导理念出现在他的其他设计中。

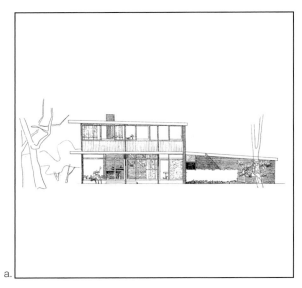
a.

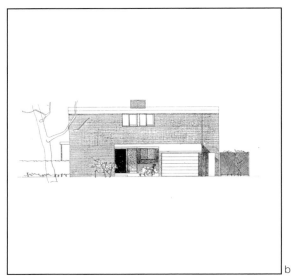
b.

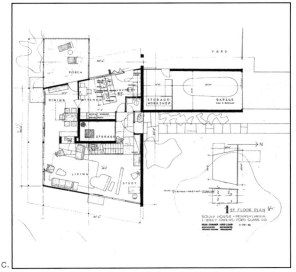
c.

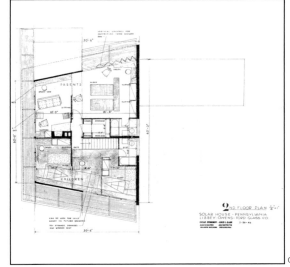
d.

In 1946, the Libbey Owens Ford Glass Company sponsored a study to investigate the feasibility of thermopane glass as it involves building materials technology, lighting engineering, and particularly architectural implementation. The company invited 49 architects to submit designs. The Pennsylvania Solar House project included new, future-oriented, as well as already-known elements. The most noteworthy characteristic is the thick brick wall facing the north, with its small bathroom windows, and the openings in the entrance area. The east and west facades are conically oriented toward the sun; as does the south side, they feature continuous glass fronts. In order to shade these huge window surfaces from direct sunlight, Kahn provided four kinds of control elements: the projecting brisesoleil with obliquely installed slats over the ground floor (cf. the Broudo House), the flat roof with projection far out beyond all sides of the upper floor (cf. Carver Court), as well as two new developments by Kahn. These were the sliding folding wall in front of the west window, and the horizontally sliding wooden panels located at the interior of the windows (ground floor and east and south facade). Although folding shutters do not reappear in any of Kahn's later designs, the sliding panels do show up in modified, further-developed form in other of his designs as guiding concepts.

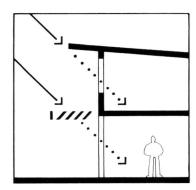

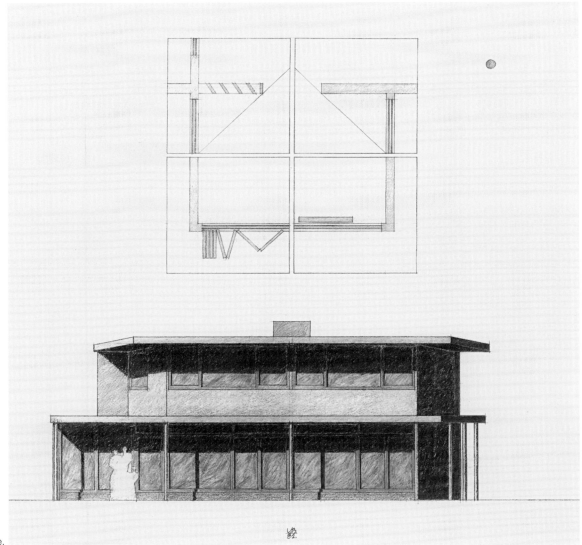

a. 东立面图
b. 北立面图
c. 底层平面图
d. 顶层平面图
e. 分析草图

a. **East elevation**
b. **North elevation**
c. **Ground floor**
d. **Top floor**
e. **Analytical drawing**

菲利浦·Q. 罗奇住宅
1947—1949；建成
康舍霍肯，
宾夕法尼亚州
与奥斯卡·格雷戈里·斯托诺罗夫合作完成

PHILLIP Q. ROCHE HOUSE
1947 – 1949; BUILT
CONSHOHOKEN,
PENNSYLVANIA WITH OSCAR
GREGORY STONOROV

乍一看，可能并不会认为这个独户住宅是出自路易斯·I. 康之手，然而这个住宅是他在探索鲍扎体系和现代建筑之间的一个重要里程碑。

在罗奇住宅中，墙体被重新诠释为内部与外部之间的一个过滤器，形式上表现为最大限度透明的薄膜一样的单层玻璃。在对称的坡屋顶下，横墙和窗户跨过内外界线，部分在内，部分在外，这些墙体之间的屋顶更强调了内在的建筑理念。此处，康尝试了五种不同的洞口，让光进入内部空间，其中包括入口区域上方的一个天窗。值得注意的还有偏离了住宅主体几何形的巨大壁炉。除了日间的自然光（阳光），夜晚的自然光（火光）也是康一生都在使用的主题。炉火作为一个原始的要素，它的实体与心理内涵一直吸引着路易斯·I. 康。

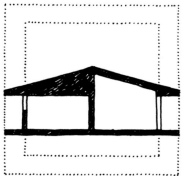

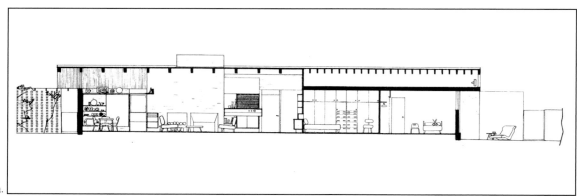

a.

b.

At first glance, one would not expect this single-family dwelling to have been designed by Louis I. Kahn and yet this project marks an important milestone in the architect's search along his way between Ecole des Beaux-Arts and modern architecture.

In the Roche House, the wall experiences a new interpretation as a filter between inside and out, in the form of a single-layer glass membrane with a maximum of transparency. Under a symmetrical gable roof, the cross walls and windows are installed partly on an outer and partly on an inner boundary line, with emplacement of the roof between these walls as reinforcement of the intended architectural concept. Here, Kahn experiments with five different types of openings to admit light into interior space including a skylight over the entrance area. Noteworthy is also the huge fireplace, displaced from the main geometry of the house. In addition to the natural light of the day (the sun), natural light in the night (fire) is a topic which occupied Kahn all his life. Fire as primeval element, with all its physical and psychological connotations, never ceased to fascinate Louis I. Kahn.

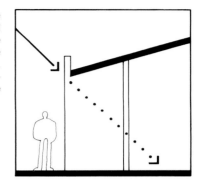

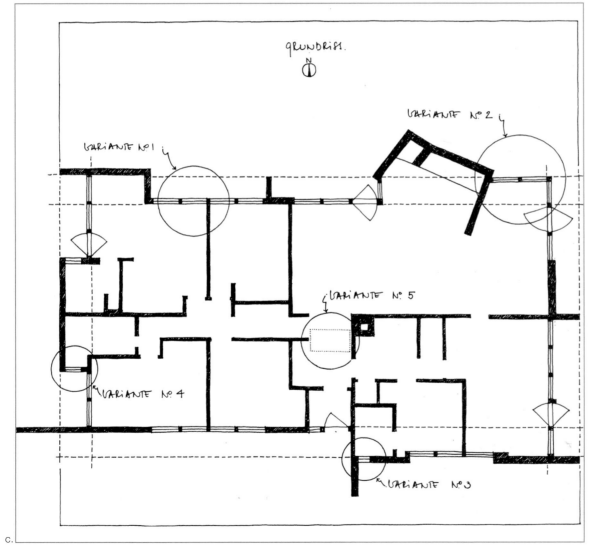

a. 东西向剖面图
b. 北向视角
c. 分析草图

a. **East-west section**
b. **North view**
c. **Analytical drawing**

67

莫顿·韦斯住宅
1947—1950；建成
东诺里斯敦，
宾夕法尼亚州

MORTON WEISS HOUSE
1947 – 1950; BUILT
EAST NORRISTOWN
TOWNSHIP,
PENNSYLVANIA

韦斯住宅或许是路易斯·I. 康早期作品中最有趣的一个，也是最后一个。从 1948—1954 年的犹太社区中心和 1951—1953 年的耶鲁大学美术馆开始，没有经过明显的过渡期，康的作品便进入了晚期。

韦斯住宅项目中充满了康在之前 24 年积累的思想与洞见。他与现代主义的纠缠似乎已经化解。楼层平面按照功能分为三个部分：车库、起居/用餐空间及卧室。原石墙体令人想起密斯·凡·德·罗早期的设计中，横墙像翅膀一样延伸到景观中。住宅的核心是一个巨大的下沉式壁炉。康第一次设计并绘制了一面壁画，作为室内重要的组成部分。在屋顶南侧位于起居室和卧室之间的开口以及跨在西侧上方的井字梁，令人想起弗兰克·劳埃德·赖特的作品。"V"形的屋顶似乎与 1930 年柯布西耶设计的埃尔祖里斯住宅几乎完全相同。这个"V"形能使透进来的光尽量远地照进房子里。然而，最主要的设计理念则是在两层通高的起居室中体现出来。康的遮阳板第一次与横向木百叶共同出现。一个由混凝土浇筑的低温供热系统，用来平衡巨大的悬挑。上部构造将非直射阳光整合。屋檐远远地挑出，减小阳光和天气变化对立面的影

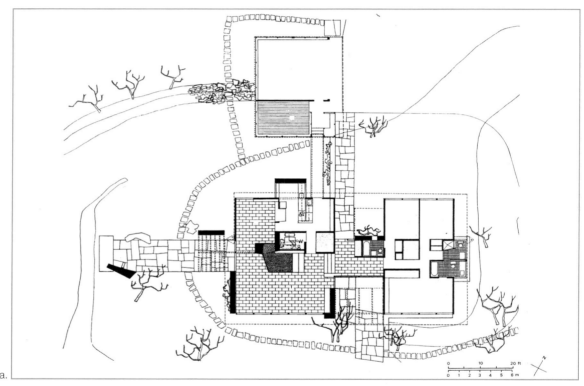

a.

b.

The Weiss House is perhaps the most interesting project as well as the last from Louis I. Kahn's early work. Beginning with the Jewish Community Center of 1948 - 1954 and the Yale University Art Gallery of 1951 - 1953, Kahn's work, without obvious transition, moves into its late phase.

The Weiss project is packed full of ideas and insights stored up from the previous 24 years. Involvement with Modernism appears to have been resolved. The floor plan is functionally divided into three wings: garage, living/dining area, and bedrooms. The rough-masonry walls remind us of early designs by Mies van der Rohe with cross walls which reach up out of the landscape like wings. The core of the house is a mighty, sunken fireplace. For the first time, Kahn makes a design and drawing for a wall painting as integral part of the interior. The cutout in the roof on the south side, between the living room and the bedrooms as well as the pergola beam spanned on the west side call to mind the work of Frank Lloyd Wright. The V-shaped roof appears almost identical to that of the Maison Errazuris designed by Le Corbusier in 1930. This V form allows light to penetrate as far as the longitudinal axis of the house. The primary design concepts, however, are realized in the large, two-story living room. For the first time, Kahn's brisesoleils appear with transversal wooden louvers. A low-temperature heating system, cast in concrete, serves as counterweight for the extreme overhang. Indi-

c.

d.

a. 平面图
b. 南立面
c. 起居室
d. 说明性草图

a. Plan
b. South facade
c. Living room
d. Explanatory drawing

响。屋顶的橡子作为可见的结构构件在室内暴露出来；与立面的支撑结构，共同形成一个空间网格。高高的檐口饰带形成了屋顶清晰的水平边框。然而，更有趣也更重要的是那些窗户中垂直滑动的木质嵌板。康在三张草图中详细阐释了他的理念。外部右侧窗户安装玻璃的区域超过了两层高。这些嵌板能按照居住者的愿望，通过滑动将室内外之间的视野连通或分隔。总共有14种不同的排列方式来调节内部空间与外部景观和天空之间的关系。因此内表面形成了巨大的景框，其中的画面可以按照自己的愿望来选择和组合。两种极端的情况：在白天完全打开，或者在夜里全部关闭来获得安全感。然而，为了实现这个绝妙的想法，技术开支非常巨大的（参阅细部图纸）。事实上，没过多久这些木板就变形了，这个构件也不能用了。不过，康还是在后来的两个项目中应用了这个理念。

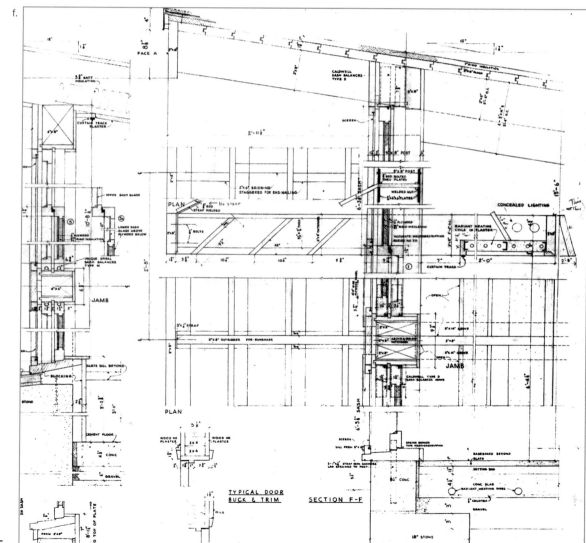

e. 剖面图
f. 细部草图
g. 分析草图
h. 南立面细部

e. Section
f. Detail drawing
g. Analytical drawing
h. South facade detail

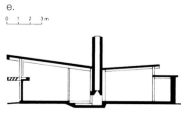

rect lighting is integrated above. A far-projecting canopy protects the facade from sunlight and weather. The roof rafters are visible throughout the interior as visible structural elements; together with the facade supports, they form a spatial grid. A high eaves fascia provides a pronounced horizontal border to the roof. More important and more interesting, however, are the vertically sliding wooden panels in the windows. Kahn elaborates on his concept here in three drawings. The window area at the outside right is glazed over two stories. As the occupant desires, the panels may be shifted to optically connect or divide the interior from the exterior. There is a total of 14 variations for arranging the interior in its relationship to landscape and sky. The interior facade therefore becomes a huge picture frame in which picture sections can be selected and combined at will. Extreme variations: entirely open during the day, and closed completely for protection at night. The technical expenditure for this wonderful idea, however, is enormous (see detailed plan). It did not take long, in fact, until the wood warped, and the mechanism no longer functioned. Kahn, nevertheless, implemented this concept in two later projects.

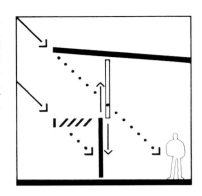

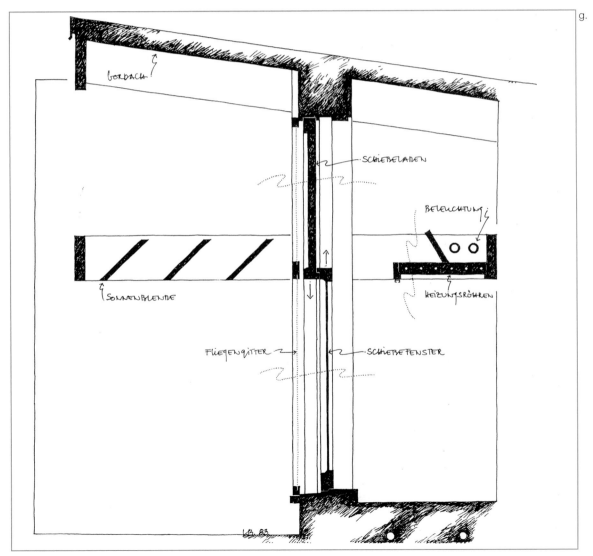

g.

h.

费城精神病医院
伯纳德·S.平克斯楼
1948—1954；建成
费城，
宾夕法尼亚州

BERNARD S. PINCUS
BUILDING,
PHILADELPHIA
PSYCHIATRIC HOSPITAL
1948 – 1954; BUILT
PHILADELPHIA,
PENNSYLVANIA

这个矩形的建筑屋顶以桁架梁支撑，桁架看起来像是落在窗户框架上。然而事实上这些大梁上的荷载是经过天棚屋顶传递到独立钢柱的。平克斯楼是康对于滑动窗构件理念的又一次尝试，此处经过调整，成为一种更简洁的技术构件。较低的一排窗户上安装了大片的门状嵌板，这些嵌板可处于两种位置：打开或关闭；不能滑动（上部的窗户是不能开启的）。然而，不可开启的门窗结构外部的木质嵌板和框架可通过一个滑轮控制系统在彼此间前后移动。

"一间能根据不同功能快速呈现出不同特质的房间，是一种精神财富，因为它能满足人类天生对于变化的渴望。将百叶和门打开，休憩区与外部门廊融为一体。百叶和门关闭，在夜晚病人们会看到一面有安全感的嵌板墙。"（14）因此形成了三种基本形式：白天（关上门），棋盘样的形式；打开门，连续通透的形式；在夜晚（木质嵌板放下），形成一面中等高度的墙体。

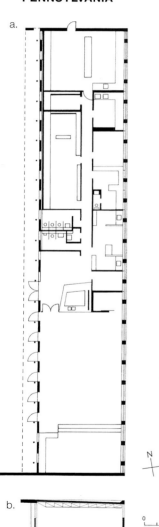

This oblong building is spanned by trussed girders which appear structurally to rest on the window frames. These girders, however, actually transfer their loads through the canopy roof to freestanding steel columns. The Pincus Building is a further attempt by Kahn to realize the concept of sliding window sections, with modification here in the form of simpler technical implementation. The lower row of windows is provided with large-leaf, door-type panels which have either of two positions: open or closed; sliding is not possible (the window installed above is non-opening). The wooden panels and frames outside the non-opening door and window structures, however, can be moved back and forth in relation to each other, by means of a pulley and cable control system.

"A room which can quickly assume a different aspect for different functions is a psychological asset, because it satisfies a natural craving for variety. With shutters up and doors open, the recreation area merges with the outdoor terrace. With the shutters down and the doors closed, patients see a paneled wall that gives a sense of security at night." (14) Three basic patterns therefore result: during the day (with closed doors), a chessboard pattern. With opened doors, a continuous and transparent pattern. And at night (with the wooden panels lowered), a medium-height wall.

a. 平面图
b. 剖面图
c. 西立面
d. 内部
e. 分析草图

a. Plan
b. Section
c. West facade
d. Interior
e. Analytical drawing

费城精神病医院
塞缪尔·拉德贝尔楼
1948–1954；建成
已改建
费城，
宾夕法尼亚州

SAMUEL RADBILL
BUILDING,
PHILADELPHIA
PSYCHIATRIC HOSPITAL
1948 – 1954; BUILT
ALTERED
PHILADELPHIA,
PENNSYLVANIA

四年前完成的早期设计最终替换成了一个功能性更强的"Y"形平面。此外，在这个医院的第二次加建中，康尝试将原来的平克斯楼与第一次加建结合起来，使三个部分形成一个统一的整体。这个设计是基于一个混凝土框架结构体系，沿中轴线上有一排柱子。沿楼板通长的带形窗安装在板岩饰面墙体的上部。采用上悬窗通风。为了遮挡照射在带形窗上的直射阳光，东、西两侧的混凝土楼板从外墙向外悬挑出来。在最初的设计中，康已经采用了遮阳板的理念；而最终的解决方案在形式和结构上更加有趣。空心木块悬在挑出的混凝土板中。这些"光过滤器"的尺寸大小，使得在一天中的某些特定时间，只有一小部分阳光能够落在墙面上。最终成果是在阴暗的背景（墙板和玻璃）上形成了绝妙的几何形光影图案。有一处发生了理念冲突，即上层的遮阳板逐渐变窄，而每一层的带形窗却都是一样高的。西南立面上的窗没有遮阳。

a. 遮阳板细部
b. 剖面图
c. 平面图（一层）
d. 东北立面

a. Detail of brisesoleil
b. Section
c. Plan (1st upper story)
d. Northeast facade

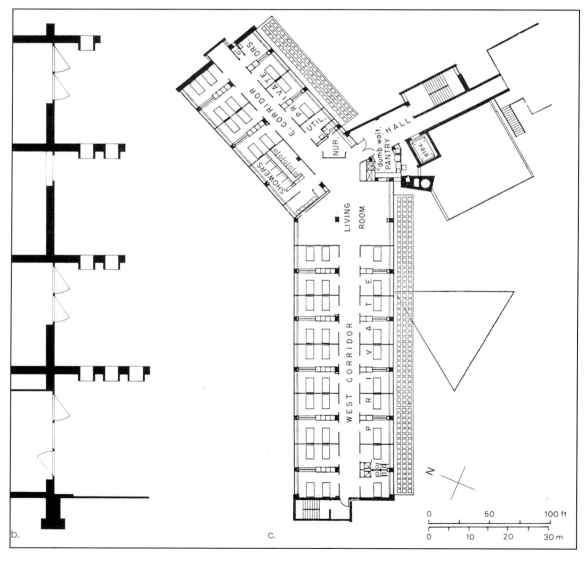

The preliminary design prepared four years earlier is replaced here by a functionally superior Y solution. Moreover, Kahn now attempts, in this second hospital addition, to unite the original Pincus Building with its first addition to form a unified ensemble of three parts. The design is based on a concrete-skeleton construction with a row of columns arranged along the central axis. The strip windows, extending entirely to the ceiling, are installed above slate-paneled apron walls. Ventilation is provided by pivoting open the top-hung windows from the bottom outward. To protect the strip windows from direct sunlight, the concrete floor slabs project beyond the plane of the facades on the east and west sides. In his preliminary design, Kahn had already included the concept of brisesoleils; the eventual solution, however, was formally and constructively more interesting. The projecting concrete slabs support suspended hollow wooden cubes. These "light filters" are dimensioned such that, at particular times of day, only a fraction of the incident sunlight will fall onto the facade. The result is a wonderful geometric light pattern formed on the dark background (slate and glass). A conceptional inconsistency are the brisesoleils which taper toward the top, although the strip windows have the same height throughout all floors. The windows of the south-west facade are not sunshaded.

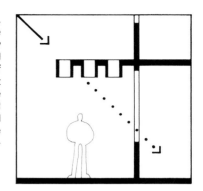

d.

耶鲁大学美术馆
1951—1953；建成
纽黑文，
康涅狄格州

YALE UNIVERSITY ART
GALLERY
1951 – 1953; BUILT
NEW HAVEN,
CONNECTICUT

耶鲁大学美术馆是路易斯·I. 康在他的家乡费城之外的第一个重要作品。清晰有力的体量使康在耶鲁的这个设计格外引人注目。由于对空间布局灵活性的需求，康设计了两个大空间单元。单元之间由一部旋转楼梯和一个中心服务区分隔开（"服务空间与被服务空间"）。为了呼应现存建筑的新哥特式立面，康设计了一面仅装有四条滴水槽的巨大砖墙。入口部分后退，以此来强调新与旧之间的过渡。北侧和东侧是整面的玻璃幕墙。"封闭"与"开敞"的对比格外强烈。强有力的石墙与整面的玻璃幕墙都是定义街道空间的元素：它们代表了路易斯·I. 康的建筑语言的全新发展。

耶鲁大学美术馆最初设计中大面积的窗户存在很大的问题，不仅牺牲了私密空间，而且很难对光线进行控制。因此，大部分的玻璃墙面现在都装上了百叶。去往各个房间，需取道一个独立又封闭的圆形楼梯塔。康将这部楼梯设计成了一个内接三角形。此处，康将四面体结构屋顶作为主题元素。一圈玻璃砖收束了圆柱体，并使光线照射进来。三角形结构的混凝土板成了"修光片"，将光线向下反射。三角形的内

a.

The Yale University Art Gallery is Louis I. Kahn's first major contract outside his home town of Philadelphia. Clarity and power distinguish Kahn's Yale design. Owing to required flexibility in the space assignment plan, Kahn designed units consisting of two large rooms each. These units are divided by a cylindrical stairwell and a central service block ("serving spaces and served spaces"). In response to the existing neo-Gothic facade, Kahn executed a great brick wall structured only by four horizontally installed water-drip molds. The facade of the entrance section is set back in order to highlight the transition between old and new. The north and east facades are entirely glazed. The difference between, "closed" and "open" could not be more pronounced. The mighty masonry wall as element for definition of street space, as well as complete glazing of the facade: these represent radically new developments in the architectural language of Louis I. Kahn.

The large windows originally provided at Yale proved to be problematic owing to the sacrifice of wall space and the difficulty of light regulation. As a result, they are for the most part now covered by panels.

Access to the rooms is via a freestanding, intrinsically enclosed, round stair tower. Kahn designed the stair spiral as an inscribed triangle. The structure of the tetrahedron-shaped roof construction is given thematic treatment by Kahn here. A ring of glass bricks forms the end of the cylinder and allows natural light to enter. The con-

a. 西南立面（草图，晚期版本）
b. 东北立面

a. **Southwest elevation (sketch, late version)**
b. **Northeast facade**

b.

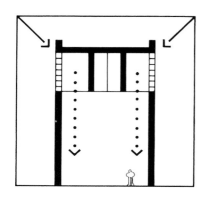

部空间处于阴影中；圆柱体和三角形形成了鲜明的对比。

四面体混凝土楼板中有纵向连续的空腔，容纳了通风管和射灯。从而三维的楼板结构与平面的墙体和窗户形成对比。楼板边缘突出出来，尽可能优美地成为玻璃的封边及窗挺。遮阳用的织物幔帘挂在最后一排四面体托梁（向内倾斜）与窗之间，从地板一直延伸到突出部分的下边缘。康是第一次使用这种遮阳构件，也是最后一次；它被证明没有什么效果。这个位置后来装上了宽宽的白色窗帘。

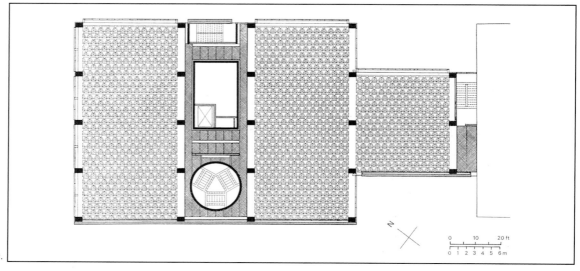

c.

d.

e.

c. 标准层平面图
d. 楼梯
e. 圆柱形楼梯间顶部的收束

c. Typical plan
d. Stairway
e. Top closure of stairway cylinder

crete plates of the triangular construction serve as "light blades" which deflect the light downward. The interior space of the triangle lies in shadow; the cylinder and the triangular form sharply contrast.

The tetrahedral concrete ceiling slab has continuous hollow spaces in the longitudinal direction which accommodate ventilation ducts and spotlights. The resulting three-dimensionally modulated ceiling structure represents a stark contrast to the wall and window sections, which remain plane. The ceiling edge is executed as an underprop in order to implement as delicately as possible the elements of the glass envelope and its muntins. Hanging textile strips are provided as sunshades between the last tetrahedral joist (canted toward the inside) and the window extending from the floor up to the lower edge of the projection. These strips can be adjusted by rotating and sliding. Kahn proposed this sunshade element here for the first and last time: it proved ineffective. Wide white curtains were installed in their place.

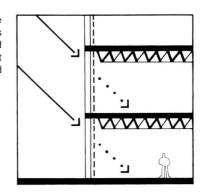

f.

g.

h.

f. 楼板结构
g. 西北立面
h. 窗户细部

f. Floor construction
g. Northwest facade
h. Window detail

伦纳德·弗鲁赫特住宅
1951—1954；未建成
费城，
宾夕法尼亚州

LEONARD FRUCHTER
HOUSE
1951 – 1954; UNBUILT
PHILADELPHIA,
PENNSYLVANIA

我认为，弗鲁赫特住宅是另一种有趣的设计：康第一次用清晰明确的几何形进行设计，他构建了一个中心，与周边的房间连接，形成放射状。星形的楼层平面由属于基本图形的圆形、三角形和正方形组成，根据特定房间的功能进行组织。三角形的内部空间用作入口区域及通往其他房间的通道。两层高的起居空间标高略低，位于住宅的第一侧翼之中。第二侧翼中容纳了厨房、餐厅和客房。卧室位于第三侧翼。圆柱形火炉间及壁炉（朝向起居室）的位置偏离了住宅中心。弗鲁赫特住宅是康使用竖向木质滑动嵌板理念的最后一次尝试。与韦斯住宅和平克斯楼相反，在弗鲁赫特住宅中完全没有遮阳构件。滑动百叶和窗户与外墙面齐平，来强调立方体体量的纯净。奇怪的是，康在起居室大空间上方的大跨度位置使用了木质构件形成的拉索状结构。基地边缘车库的位置非常实用。

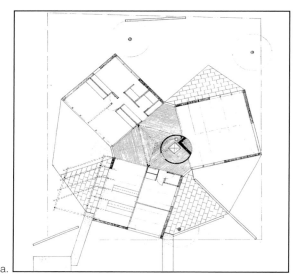

a.

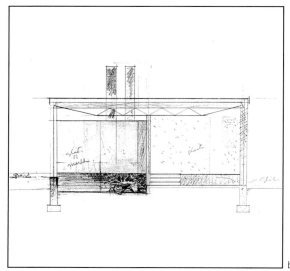

b.

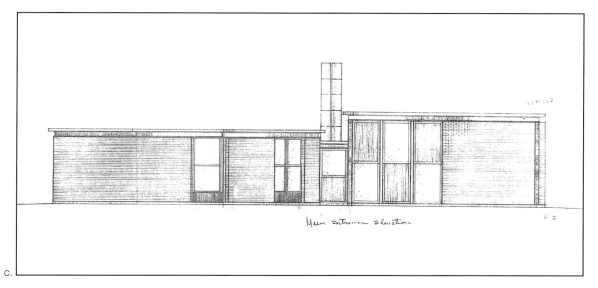

c.

In my opinion, the Fruchter House represents an exceptionally interesting design: for the first time, Kahn has designed a project with definitely clear geometric figures, one which features a center providing radial access to peripheral rooms. The star-shaped floor plan is composed of the basic forms circle, triangle, and square, and it is arranged according to the functions of the particular rooms. The triangular interior room serves as entrance area and access to the remaining rooms. The two-story living area, on a slightly lower level, is accommodated in the first wing of the house. The second wing contains the kitchen, dining room, and guest rooms. The bedrooms are in the third. The cylinder-shaped furnace room with its part of the fireplace (oriented to the living room) has been located away from the central area of the house. The Fruchter House is Kahn's last design which attempts to implement the concept of vertically sliding wooden panels. Contrary to the Weiss House and the Pincus Building, there are no sunshade elements at all in the Fruchter House. The sliding shutters and the windows have been incorporated flush into the facade, in order to emphasize the clarity of the cubic construction. It is curious that Kahn proposed to structurally span the large dimensions of the living room with the aid of a wooden construction featuring a cable under tension. The location of the garage at the edge of the property is highly practical.

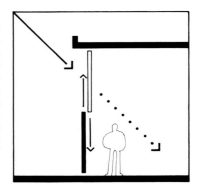

a. 平面图
b. 剖面图
c. 东北立面图
d. 分析草图

a. Plan
b. Section
c. Northeast elevation
d. Analytical drawing

美国劳工联合会
医疗服务楼
1954—1957；建成
1973 年拆除
费城，宾夕法尼亚州

AMERICAN FEDERATION
OF LABOR MEDICAL
SERVICES BUILDING
1954 – 1957; BUILT
DEMOLISHED 1973
PHILADELPHIA,
PENNSYLVANIA

对内部空间灵活性的要求使得康设计的这幢建筑与耶鲁大学美术馆有相似之处。通风管道收纳在空腹托梁的凹槽中。在这个设计中，混凝土楼板在立面上呈现为薄板的形式。窄条形的实墙和玻璃交替布置。第一眼看上去，这个设计表达了一个清晰的设计理念；然而实际上，它出现了很多不同的概念问题，这些问题并没有得到满意的解决。在东西立面上，空腹托梁位于垂直的窗户和墙内侧。康没有在立面上完全展示中间楼板结构的力量，也没有化解空腹托梁的结构高度（加上楼板厚度）与正立面表皮之间的冲突，例如，在楼板边缘出挑的情况下，康漫不经心地将北立面的空腹托梁向内移动了一个柱子的宽度。托梁被卡在左边和右边的柱子之间。除此之外，这些梁在两层通高的前厅中没有任何结构功能，而完全是为了装饰。仅仅通过挂上窗帘来获得遮阳和私密性。可以肯定，康在后来设计的耶鲁英国艺术研究中心时才解决了以上列出的问题。

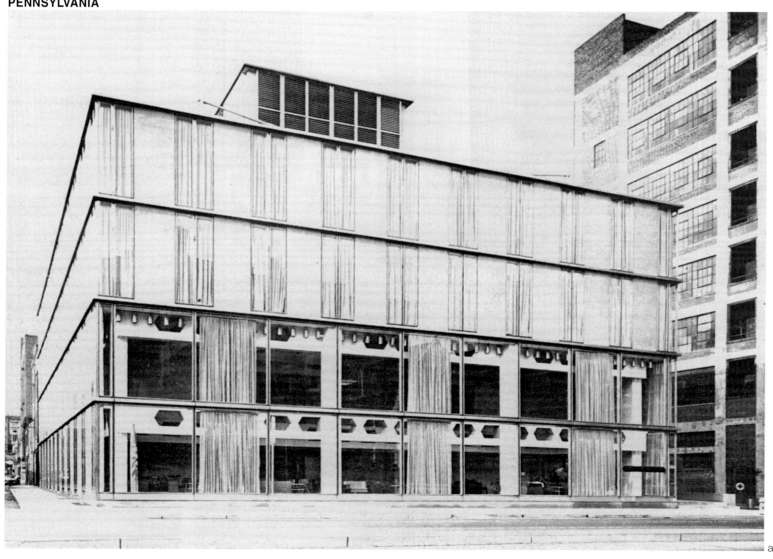

a.

The flexibility required in the interior space of this building prompted a design from Louis I. Kahn similar to the Yale University Art Gallery. The ventilation ducts are accommodated in the recesses of the Vierendeel joists. In this design, the concrete floor slabs appear in the form of thin underprops in the facade. Cross walls and glass are alternately installed in narrow strips. This design may appear at first glance to represent a clear concept; in reality, however, it presented many and various conceptional problems which were never satisfactorily solved here. At the east and west facades, the Vierendeel beams run flush behind the vertically extended window and wall sections. Instead of revealing the total strength of the intermediate floor-slab construction in the facade, or of solving the conflict between the construction height of the Vierendeel joists (plus floor-slab thickness) and the facade skin in front, e.g., by providing additional projection of the floor-slab edge Kahn halfheartedly moved the Vierendeel beams one column width back in the north facade. The joist itself is "stuck" to the left and right of the columns. In addition, these beams in the two-story foyer have no structural function and were provided strictly for decoration. Sunshade and privacy are provided only in the form of curtains. Kahn solved the problems outlined above only later exceptionally, to be sure, in the Yale Center for British Arts and Studies.

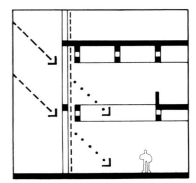

a. 北立面图
b. 前厅
c. 二层平面图

a. North facade
b. Foyer
c. First story

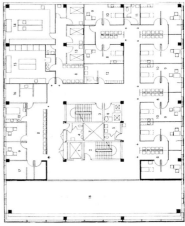

犹太社区中心
1954–1959；
浴室及日间营地建成
尤因镇，
特伦顿，新泽西州

JEWISH COMMUNITY
CENTER
1954 – 1959; BATH HOUSE
AND DAY CAMP BUILT
EWING TOWNSHIP,
TRENTON, NEW JERSEY

在整个社区中心建筑群的原始设计中，只有浴室、水池和日间营地实际建成。康中心对称的浴室设计中最鲜明的特征，是它中心对称的形式。内庭院周围有四个带金字塔状屋顶的构筑物：一个用来售票，两个是更衣室，最后一个是开放的台阶式入口。十二根巨大的空心柱支撑着屋顶。康第一次将天窗作为整个设计中必不可少的部分，但并不算特别成功，这在后面我们会看到。正如我们在罗奇住宅设计中所见，墙的二元性在此处成为一个核心问题。在售票处，墙在"内部"；而在更衣室中，则有三个侧面是在"外部"。屋顶边缘全部沿着巨大柱墩的中轴线。墙向外偏移，这个偏移不无问题。一方面，由于光会从一侧照进来，模糊了中央天窗的概念。另一方面，它将卫生洁具暴露在天空下。对光线的控制仅在售票处和楼梯间出口处算是成功的，因为这些墙位于金字塔形屋顶边缘以内，或是完全省去了。然而，康写道："在特伦顿浴室完成后，我不再需要从其他建筑师那里获得灵感。"（15）

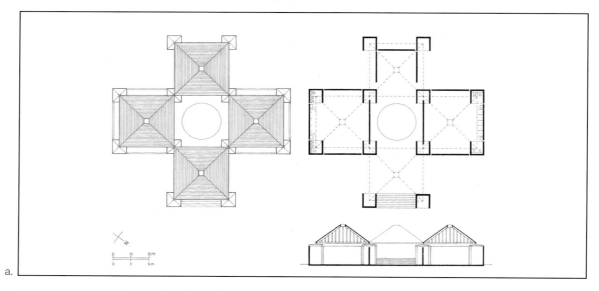

a.

b.

a. 屋顶平面图；平面图；剖面图
b. 西南视角
c. 分析草图
d. 台阶

a. Top view of roof; plan; section
b. Southwest view
c. Analytical drawing
d. Stairway

Of the entire originally planned community center complex, only the bath house, pool, and Day Camp were actually built. Kahn's design for the bath house is distinguished by its central symmetry. The interior courtyard is surrounded by four structures covered by pyramid-shaped roofs: one room for ticket sales, two changing rooms, and the open stairway entrance. Twelve massive hollow columns support the roofs. For the first time, Kahn designs a skylight as an integral part of a structure but not entirely successfully, as we shall see. Duality with walls becomes a central matter here, as we know them from Kahn's design for the Roche House. For the ticket-sales building, the walls are "inside"; in the changing rooms, they are "outside" on three sides. The roof edges run in all cases along the central axis of the powerful pillars. This displacement of the walls to the outside is not without difficulty. For one, it confuses the concept of the central skylights, due to light entering from the side. For another, it exposes the sanitary fittings to the elements. Control of light is successful here only in the ticket-sales building and in the covered stairway exit where the walls stand inside the edges of the pyramidal roof, or were left out altogether. Nevertheless, Kahn writes: "After the completion of the Trenton Bath House, I never had to look to another architect for inspiration." (15)

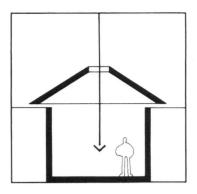

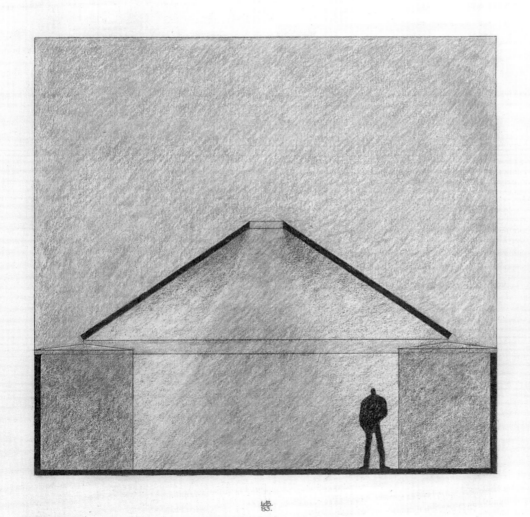

c.

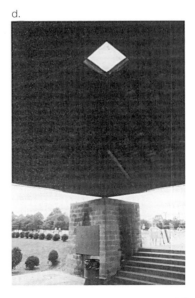

d.

劳伦斯·莫里斯住宅
1955–1958；未建成
基斯科山，纽约

LAWRENCE MORRIS
HOUSE
1955 – 1958; UNBUILT
MT. KISCO, NEW YORK

康在耶鲁大学美术馆、弗鲁赫特住宅以及特伦顿浴室中严格的几何与形式原则在这个设计中似乎消失了。除了西侧的卧室部分，楼层平面和立面轮廓均呈现出明显的无序性。巨大的垂直墙面令人想起坐落于山顶上的中世纪城堡形象。然而，通过从顶部观察模型，能看到清晰的设计意图，即通过改变房间高度使其符合每个房间自己的特性。莫里斯住宅的建筑理念是以采用隔板墙为基础的，隔板像这样一个挨着一个布置，留出一道道垂直的采光狭口，就像垂直于玻璃幕墙的支柱。方案东侧部分独立的一组隔板可以理解为仅在有限程度上起到遮阳作用：从根本上讲，它代表了一种以戏剧化的垂直性融合以及强调内部和外部概念的元素。在康的这个工作阶段，壁龛般的窗户顶部仍是开敞的，形成进一步发展的基础。就这一点来说，位于罗切斯特的第一唯一神教堂及学校与之相比，其墙体和窗户就形成了双层表皮。

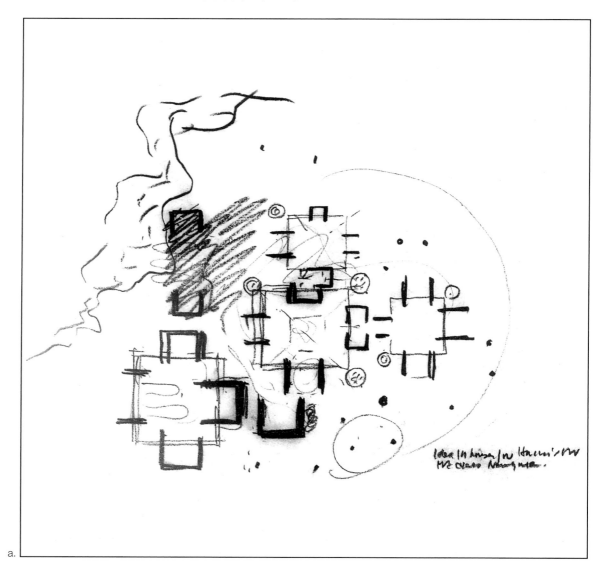

a.

The severe geometry and formal discipline which Kahn exhibited in the Yale University Art Gallery, in the Fruchter House, and in the Trenton Bath House appear to have dissipated here. With the exception of the bedroom section on the west side, the floor plan and the silhouettes of the facades are characterized by an apparent lack of order. The massive vertical walls evoke images of a medieval castle on the peak of a mountain. Upon examination of the model from above, however, the intention becomes apparent of according each room its own identity through variation of room height. The architectural concept of the Norris House is based on the employment of cross walls which, arranged next to each other as they are here, leave open vertical light slots which stand as piers at right angles to the glass facade. The freestanding bundle of cross walls in the east section of the project should be interpreted to only a limited degree as serving for sunshade: primarily, it represents an element which by virtue of its dramatic verticality amalgamates and emphasizes the concepts of interior and exterior. The window niches here, still open to the top at this stage of Kahn's work, form the basis for further development. In this context, compare with the First Unitarian Church and School in Rochester, for which the wall and the window have become double-layered.

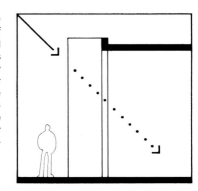

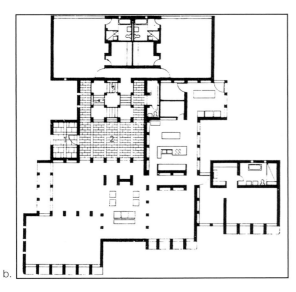

a. 草图（晚期版本）
b. 平面图
c. 模型顶视图
d. 模型西南立面

a. Sketch (late version)
b. Plan
c. Top view of model
d. Southwest facade of model

华盛顿大学图书馆竞赛方案
1956；未建成
圣路易斯，密苏里州

WASHINGTON
UNIVERSITY LIBRARY
COMPETITION
1956; UNBUILT
ST. LOUIS, MISSOURI

路易斯·I. 康的华盛顿大学图书馆设计是在一次设计竞赛中提交成果的一部分。为了将新建建筑与现有部分整合在一起，康采用了阶梯式的外观，令人联想起通灵塔。由于形式上的切角，形成了十字形的楼层平面。为了使光线充分射入相当深远的建筑内部，康在表皮内侧每两层设置了两层通高的空间。而且，康没有让光线从中心对称结构的中间区域顶部射入。为了遮挡窗口的直射阳光，康设置了遮阳板，这些遮阳板相互之间略微错落，安装在南面。在西侧，康设计了由"V"形金属构件组成的垂直遮阳板。然而奇怪的是，他没有在东侧设置任何遮阳构件。康后来继续采用通灵塔的形式进行设计，但与圣路易斯的项目不同，他在后续项目中考虑了顶部采光的可能性：参阅联合神学研究院公共图书馆和埃克塞特学院图书馆的项目分析。

a.

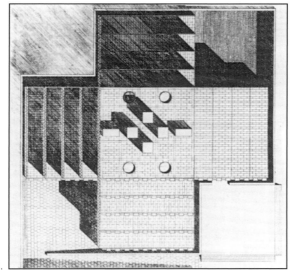
b1.

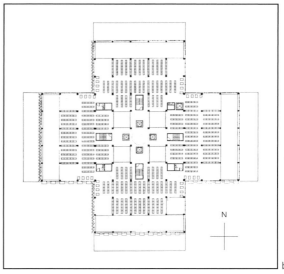
b2.

Louis I. Kahn's design for the Washington University Library was submitted as part of a competition. In order to integrate the new construction into the existing facilities, Kahn provided it with a stepped exterior, reminiscent of a ziggurat. The pattern resulting from the cut-out corners lends the floor plan a cruciform. In order to allow light to penetrate sufficiently into the relatively deep building, a double-story room directly behind the facade is provided after every second story. Kahn, furthermore, did not design the middle area of the centrally symmetric construction to permit light to enter from above into the interior of the building. To shade the windows from direct sunlight, Kahn proposed brisesoleil panels, slightly staggered with respect to each other, for installation on the south side. For the west facade, Kahn planned a vertically configured light filter consisting of V-shaped metal elements. Strangely, however, he designed no elements to shade the east facade. Kahn continued to employ the ziggurat form and, unlike at St. Louis, he treated the possibility of a light source from above in later projects: see the project descriptions on the Common Library for Graduate Theological Union (GTU), as well as on Phillips Exeter Library.

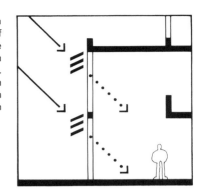

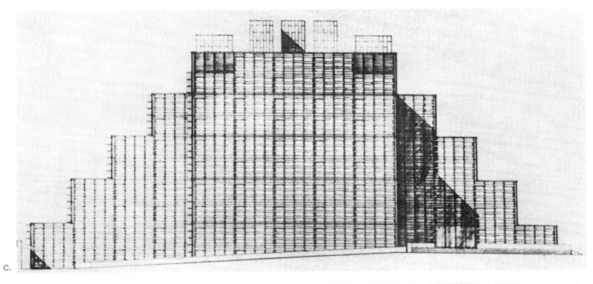
c.

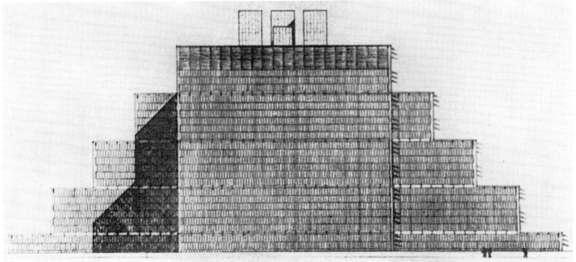
d.

a. 南立面草图（晚期版本）
b1./b2. 屋顶顶视图；平面图
c. 南立面图
d. 西立面图

a. South elevation sketch (late version)
b1./b2. Top view of roof; plan
c. South elevation
d. West elevation

伊莱恩·考克斯·克莱弗住宅
1957–1962；建成
樱桃山，
新泽西州

ELAINE COX CLEVER
HOUSE
1957 – 1962; BUILT
CHERRY HILL,
NEW JERSEY

a. 平面图
b. 剖面图
c. 南立面
d. 起居室

a. Plan
b. Section
c. South facade
d. Living room

在看过特伦顿浴室之后，克莱弗一家委托路易斯·I.康为他们设计自宅。早期项目中的几何原则在克莱弗住宅中弱化了。此处，康在中心对称的起居室周围成组设置了"附属空间"，这些结构的尺寸是依据各自的功能确定的，也设计了金字塔形的屋顶。起居室格外宽敞，与四周的附属结构形成对比，给人留下有点不切实际的印象。起居室十字形的屋顶承托在四个巨大的"L"形混凝土体块上。屋脊向房间外远远伸出，遮挡了直射三角形窗的阳光，并使它与自然隔绝。这些资料展示了屋顶的结构特征，因为它是基于三角形的几何形式。在起居室每一个不能开启的三角形窗旁边，康都安装了木嵌板，这些嵌板可以开启，为这个高大的空间获取充裕的通风。康在后面的项目中继续了这个理念：参阅埃西里科住宅和费舍住宅的资料。尽管这些窗户不能打开，但它们在此处第一次与通风嵌板共同联合组成了一个兼顾功能与形式的整体。

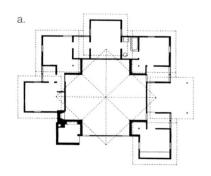

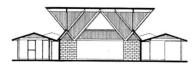

After they had seen the Bath House in Trenton, the Clever family commissioned Louis I. Kahn to design their own home. The geometrical discipline of this earlier project is moderated for the Clever House. Here, Kahn grouped "satellites" around the centrally symmetric living room: these are structures dimensioned according to their respective functions, and designed with pyramidal roofs. The living room is characterized by surprising spaciousness, in contrast to the surrounding annex constructions, which make a somewhat impractical impression. The cruciform-designed roof of the living room rests on four massive L-shaped concrete leaves of blocks. The ridge purlins, which protrude far out of the room, protect the triangular windows from direct sunlight and from the elements. The documents here depict the structural features of the roof construction, based as it is on the geometric form of the triangle. At the side of each non-opening triangular window in the living room, Kahn installed wooden panels which can be opened to provide sufficient ventilation for this high space. Kahn pursued this idea further in later projects: see the material on the Esherick House and the Fisher House. Although these windows cannot be opened, their design executed for the first time here allows them to merge with the integrated ventilation panels to form a functional and formal unity.

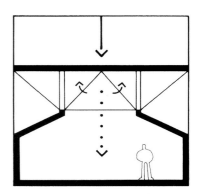

d.

阿尔弗雷德·牛顿·理查德医学和生物研究中心
1957–1965；建成
费城，
宾夕法尼亚州

ALFRED NEWTON
RICHARDS
MEDICAL RESEARCH
BUILDING AND BIOLOGY
BUILDING
1957 – 1965; BUILT
PHILADELPHIA,
PENNSYLVANIA

这个设计具有清晰的结构特征。研究所的工作空间以正方形为基础，并可以根据需要灵活地进行进一步划分。这个空间两侧的砖塔令人印象深刻，塔内容纳了通风管道和疏散楼梯。然而，实验室的功能有效性被证明存在相当大的问题。不管怎样，这些塔楼形成的建筑印象令人想起路易斯·I. 康 1928 年第一次旅欧时，在圣吉米亚诺画下的速写。混凝土框架融入砖塔之中，沉重地压在大地上。然而，由于混凝土框架的空腹支撑在边缘处变窄，给了人一种水平的轻盈感。混凝土框架上的砖砌扶手与窗户齐平。内部空间结构与外部表皮之间的相互作用在此处显得意味深长。另一方面，窗口没有遮阳：特制的蓝色玻璃被证明是不够的。后来使用者不得不在窗户内侧装上了铝箔。为了最大限度获得透明感和轻盈感，顶部的玻璃在转角处直接用硅密封胶和 L 形金属夹连接，没有窗梃。

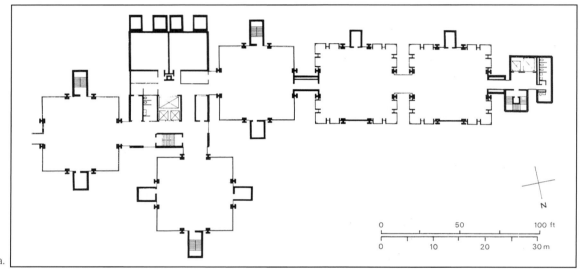

a.

b.

c.

This design is characterized by structural clarity. The working space of the institutes is organized on the basis of the square, and can be flexibly subdivided as required. This space is flanked by imposing brick towers containing supply and exhaust ventilation ducts and the emergency staircase. The functional effectiveness of the laboratory rooms, however, has proved to be a highly problematic matter. At any rate, the architectonic impression rendered by these towers calls to mind the sketches of towers made by Louis I. Kahn in San Gimignano on his first trip to Europe, in 1928. A concrete skeleton construction has been integrated into the brick towers, which ponderously weigh on the earth. The concrete skeleton, however, by virtue of Vierendeel supports tapering toward the periphery, evoke a sense of horizontal lightness. The brick parapets have been installed flush with the windows onto concrete brackets. The interplay between the interior room structure and the exterior envelope is demonstrated expressively here. On the other hand, the windows are not protected from the sun: the special blue-tinted window glass proved to be insufficient. The occupants were later forced to install aluminum foil on the inside of the windows. To achieve maximum transparency and lightness, the sheets of window glass at the top were directly connected at the corners with silicon sealing compound and L-formed metal clamps, instead of framing.

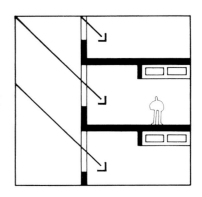

d.

a. 平面图
b. 混凝土框架
c. 实验室
d. 窗户细部

a. **Plan**
b. **Concrete skeleton**
c. **Laboratories**
d. **Window detail**

论坛报出版公司大楼
1958–1962；建成
格林斯堡，
宾夕法尼亚州

TRIBUNE REVIEW
PUBLISHIING COMPANY
BUILDING
1958 – 1962; BUILT
GREENSBURG,
PENNSYLVANIA

和特伦顿浴室一样，康在这个建筑中也采用了粗糙混凝土砌块。大跨度的预应力混凝土梁跨在外墙与服务内核之间。这些梁支撑在由相当小的石砌块砌筑的支承构件上，并在外立面上悬挑出来。这个结构体系中安装着非承重的填充墙。这些墙体上第一次出现了由路易斯·I. 康发明的窗户类型：锁孔窗。它是由莫里斯住宅的窥孔窗和现代主义的带形窗组合而成的一种新形式。鲍扎体系沉闷乏味的垂直性清晰地与现代主义建筑水平的轻盈感结合到了一起。或许罗马建筑也为康提供了灵感。他的锁孔窗在视线高度处有一个眺望室外的视野，也保留了充足的实墙以便布置家具，家具高度以上的窗户为室内提供了理想的采光。康早期木质滑动嵌板的尝试以及延伸至屋顶的窗或许都对这个窗户形式的发展有所帮助。从这里开始，康进一步发展并改良了锁孔窗，形成各种改良的形式，应用在大量不同的项目中。在论坛报出版公司大楼中，服务体块上部也采用了天窗采光。

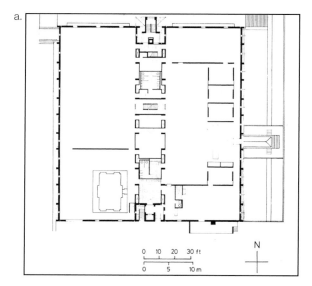
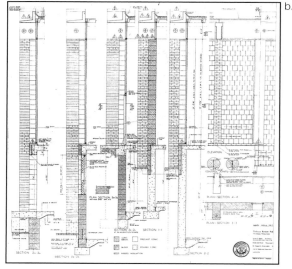

a. 底层平面图
b. 细部草图
c. 东北立面
d. 内部

a. Ground floor
b. Detail drawing
c. Northeast facade
d. Interior

As in the Trenton Bath House, Kahn uses rough concrete blocks here. Prefabricated, prestressed concrete beams bridge the long span from the exterior walls to the service core. These beams rest on supports which are built up of relatively small masonry blocks, and which appear in the exterior facade as projecting structural elements. The non-loadbearing walls are installed into this structural system. These walls feature the first appearance of a window type which is a true invention of Louis I. Kahn: the keyhole window. The loophole windows of the Morris House are amalgamated with the strip windows of Modernism to produce a new entity. The ponderous verticality of the Ecole des Beaux-Arts is apparently united with the horizontal lightness of modern architecture. Perhaps Roman architecture also served as an inspiration to Kahn. His keyhole windows allow an eyelevel field of vision toward the outside, they permit sufficient wall area to profitably furnish the floor space, and the window area above the furnishing ensures optimal illumination of the room. Kahn's earlier experiments with sliding wooden panels, and his use of windows extending directly to the ceiling, may have contributed to development of this window type. From this point on, Kahn further developed and refined his keyhole window, in various modifications and in a great variety of projects. At the Tribune Review Building, skylights are also installed above the service block.

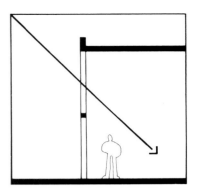

d.

罗伯特·H. 弗莱舍住宅
1959；未建成
埃尔金斯·帕克，
宾夕法尼亚州

ROBERT H. FLEISHER
HOUSE
1959; UNBUILT
ELKINS PARK,
PENNSYLVANIA

弗莱舍住宅的设计是严整的轴对称形式。为了强调中轴线，康将两个正方形体块夸张地分开，并基于此设计了低半层的起居室。然而，从外观上并不能明显看出这是一幢错层住宅。这幢建筑中再次采用了锁孔窗，锁孔窗是在论坛报出版公司大楼中首次出现，但此处对形式进行了进一步发展。半圆形令人想起古罗马砖结构。基础部分由砖石砌筑，其上的混凝土隔墙上开着半圆形的洞口。砖墙和不能开启的窗户可以装在这些隔墙上。康没有指出这些不开启的窗如何遮阳。下面的窗口可以向外眺望，也能用来通风。在基础部分，为了符合功能要求，康将洞口的形式进行了改变。露台处，康将半圆形洞口和眺望的窗口敞开；或者，他将这些洞口延伸到地面，当作门洞。入口上方，它们组合在一起，让阳光照射进来，也能由此向外眺望。康对三种基本几何形式（正方形、圆形和三角形）的使用，发展出了一种新的建筑语言：一种越来越有力量的建筑语言。多年来，他一直与现代主义相一致的努力，似乎从此结束了。

A 屋顶板支架（花园单元没有）
B 屋顶板嵌槽
C 固定窗扇或砖壁板
D 可开启的窗扇

A Ledge for roof plank (not for the garden units)
B Reglet for roofing
C Fixed window or brick panel
D Operating window slot

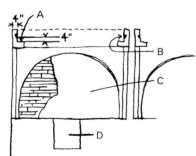
a.

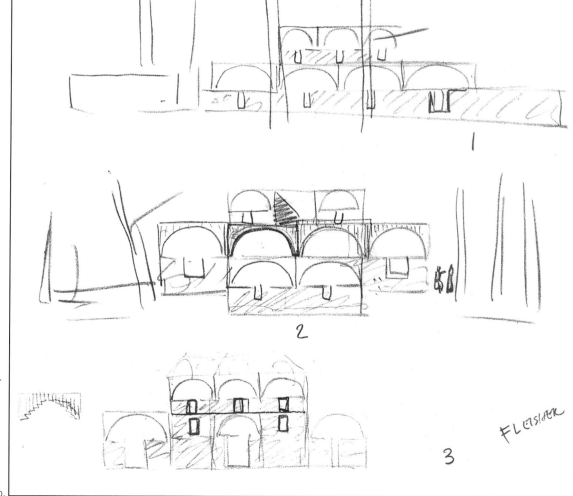
b.

The axially symmetric design of the Fleisher House is strict. In order to emphasize the central axis, Kahn planned for exaggerated separation of the two squares on the basis of which the living room lower by a half story is conceived. It is not apparent from the outside, however, that this is a split-level home. The keyhole window first employed for the Tribune Review Publishing Company Building appears here as well, but in further-developed form. The semicircles call to mind Roman brick structures. The footing consists of masonry, and the concrete cross walls above the footing feature semicircular cutouts. Brick walls or non-opening windows could be installed into these cross walls. Kahn does not indicate how such non-opening windows might be shielded from the sun. The window slots below serve for an outward view and for ventilation. Kahn varies the form of this opening in the footing area according to functional requirements. In Kahn's gazebos, the semicircles as well as the viewing slots have been left open; or, he extends them to the floor to serve as doors. Above the entrance, they combine to admit sunlight and to allow looking out. Kahn's recourse to the three basic geometrical forms (square, circle, and triangle) opens the way toward developing a new architectural language: one which will become more and more powerful. His efforts of many years to fall in with Modernism now appear to have finally ceased.

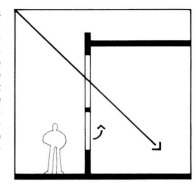

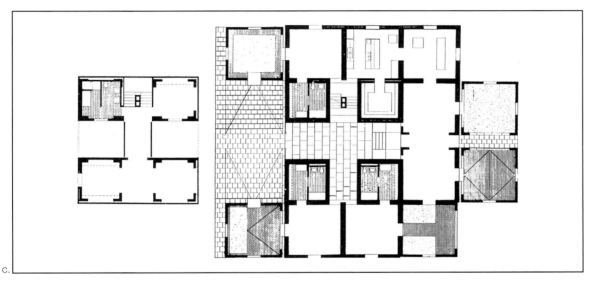

a. 说明性草图
b. 草图（晚期版本）
c. 底层和上层平面图
d. 模型

a. Explanatory sketch
b. Sketch (late version)
c. Ground floor and upper floor
d. Model

玛格丽特·埃西里科住宅
1959–1961；建成
栗子山，
费城，
宾夕法尼亚州

MARGARET ESHERICK
HOUSE
1959 – 1961; BUILT
CHESTNUT HILL,
PHILADELPHIA,
PENNSYLVANIA

在康所有的项目中，我在埃西里科住宅的研究上投入了最多的精力。这个住宅有两个最重要的特质。首先是与众不同的窗户系统，通过调节通风嵌板，为"外部"与"内部"的联系或分隔提供了多种不同的可能。其次是平面图和剖面图中"服务空间与被服务空间"之间清晰的划分。

1982年，我开始按年代整理费城路易斯·I. 康档案馆保存的埃西里科住宅的原始设计草图。即使在最早的版本中，也有锁孔窗和四个层级（为了"服务空间与被服务空间"）。康清晰地将结构构件分为承重构件与非承重构件。和论坛报出版公司大楼一样，承载混凝土托梁的柱子是用混凝土砌块砌筑的，即由小构件组成的。非承重的填充墙中间安装了用于观景和通风的窗扇。不能开启的窗户装在上方。在设计过程中，康逐渐将框架结构变成了传统的砌体结构：在分析草图9和图14中能清晰地看到这种发展变化。此外，两层通高的起居室从北侧移到了南侧。康按照逻辑将楼梯整合在中间的"服务"区域（分析草图15）。巨大的实墙和窗户部分清晰地控制着各自的东向、南向和西向立面。然而，在北立面仅出于

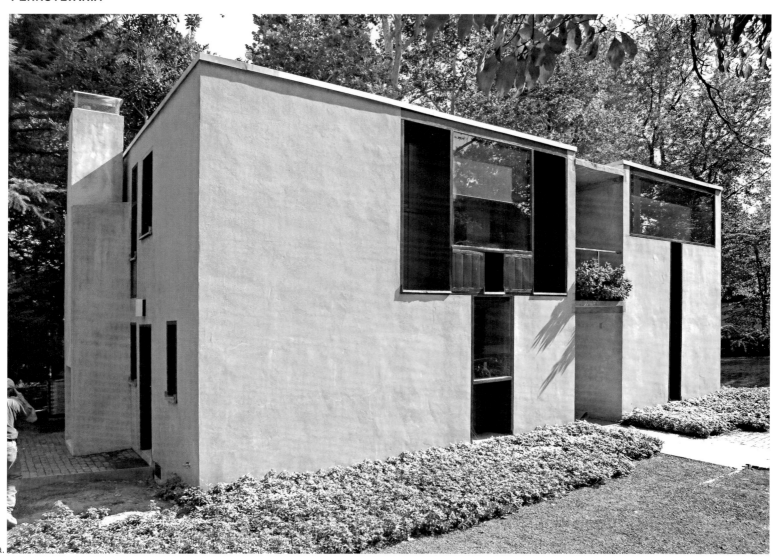

a.

Of all Kahn projects, it was the Esherick House which most extensively occupied me during my studies of his work. Two primary characteristics distinguish this house. First are the novel window systems with their various possibilities of linking or separating "outside" and "inside" by means of adjustable ventilation panels. Second is the distinct separation between "serving spaces and served spaces" in the floor plans and sections.

In 1982 I began to chronologically arrange the original Esherick sketches still existent at the Kahn Archives in Philadelphia. Even the earliest contained the keyhole windows and the four layers (for "serving spaces and served spaces"). Kahn clearly classified structural elements as loadbearing or non-loadbearing here. As with the Tribune Review Building, the columns on which the concrete joists rest are composed of concrete blocks, in turn assembled from small parts. The non-loadbearing walls are inserted there, and feature viewing and ventilation slots in the middle. Non-opening windows are installed above. During the design process, Kahn gradually transformed a skeleton structure into a conventional masonry construction: a development clearly apparent between the analytical sketches no. 9 and no. 14. The two-story living room, in addition, is shifted from the north toward the south. Kahn logically integrates the stairway into the middle, "serving" zone (analytical sketch no. 15). The large masonry-wall and window sections clearly organize the respective east, south, and

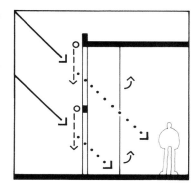

b.

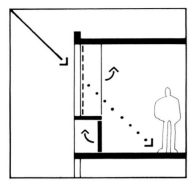

功能要求最低限度地设置了门和常规的窗户。在玛格丽特·埃西里科住宅中，康还在非开启的窗户旁边进一步改进了木质通风嵌板的细部构造（此构造的最初采用参阅克莱弗住宅）。立面上所有的玻璃（除了北立面）都是不能开启的。东南立面上安装了织物卷帘，用来遮阳。窗户旁没有玻璃的通风百叶装在垂直的壁龛中间，来获得穿堂风。通风嵌板能水平打开或关闭，通过大量变化，控制着使用者与住宅周边环境之间的视线交流。这个理念在两层通高的起居室中，与锁孔窗共同带来了奇妙的视觉体验。康将室内的西侧立面设计成了整面的窗户、书架和通风口。一旦上下布置的两个通风嵌板关闭，使用者与室外就没有视线联系了：形成了内向的室内空间。在康的设计中，书架成了一面高墙，墙后给人以安全感。然而，当将这两个通风嵌板打开，房间的特质马上发生了变化。锁孔窗使人能透过高且窄的"窥孔"望向室外，新鲜的空气也从这里进来。康本可以在通风嵌板后面装上普通的可开启玻璃窗。这样在冬天使用起来就能更自由，能像夏天一样通透，但也就让住宅在一年四季都没有什么不同了。

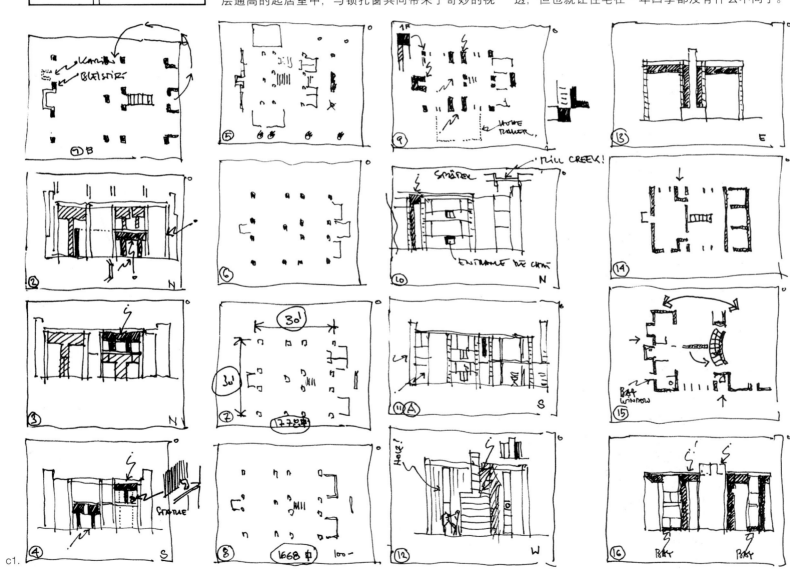

west facades. The north facade, however, is minimally provided with doors and conventional windows located according to interior functional requirements.

For M. Esherick, Kahn also further developed the detail of wooden ventilation panels next to non-opening windows (cf. the Clever House for the original employment). All glazed surfaces (except on the north facade) are non-opening. Rolling fabric shades are provided for sunshade on the outside of the southeast facade. Non-glazed ventilation shutters, next to the windows in vertical niches, allow cross-ventilation. The ventilation panels can be opened and closed horizontally to enable extensive variation in control of the occupant's visual contact with the surroundings of the house. This concept permits marvelous visual experience in conjunction with the keyhole window in the two-story living room. Kahn designed the west interior facade as an entity consisting of windows, bookcases, and ventilation opening. Once the two ventilation panels, located one above the other, are closed, the occupant has no visual connection with the outside: the interior makes an introverted impression. In Kahn's design, the bookcase becomes a high wall behind which one feels secure. Opening the two ventilation panels, however, suddenly transforms the character of the room. The keyhole window appears, one can look through the high, narrow "loophole" toward the outside and fresh air comes in. Kahn could have installed normally opening glazing behind the ventilation panels.

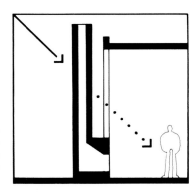

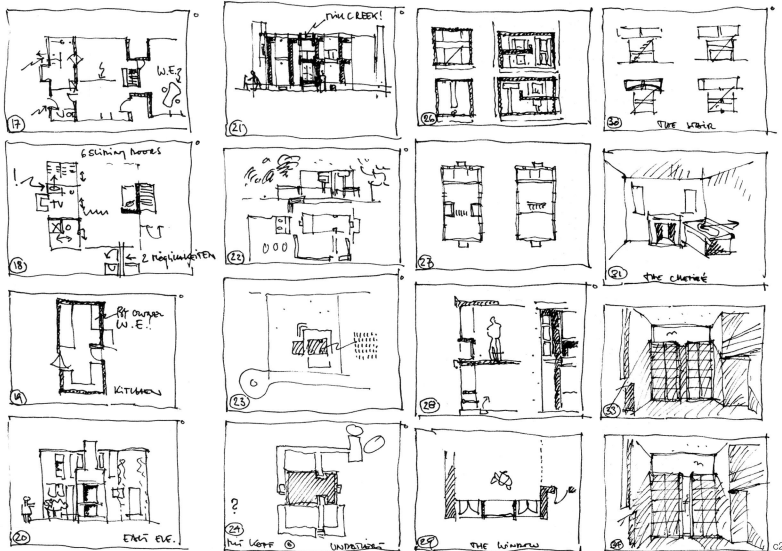

然而，康的这种处理手法，居住者会自然地跟随一年四季的步伐。冬天，可以说是蛰伏起来了；而在夏天，则将自己向着大自然敞开。卧室里，康在窗的中轴上额外装了小通风嵌板。在凉爽的季节，它们精确地调节着通风。埃西里科住宅中有两个壁炉，它们既是"夜晚的光"，也带来了温暖：一个位于起居室的南墙，另一个在楼上的一间浴室中。在那里，你可以躺在浴缸里，享受真正的炉火（分析草图18和图31）。通过一个水平移动到墙壁内的大木抽屉，浴缸变成了沙发。这幢住宅通过一套先进的热风系统采暖，然而这套系统在夏天并不能当作空调使用。路易斯·I. 康后来又为埃西里科家族设计了两个建筑：对现有埃西里科住宅的加建以及艺术家沃顿·埃西里科的工作室。

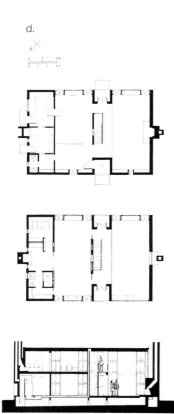

d.

e.

f.

This would have granted the additional freedom, during winter, of providing the same transparency as in summer but would have implemented a certain indifference in the house to the seasons of the year. With Kahn's solution, on the other hand, man and his dwelling naturally adjust to the annual march of the seasons. In winter, one hibernates, so to speak; in summer, one opens oneself to nature. In the bedrooms, Kahn has additionally installed small, supplementary ventilation panels at the center axis of the windows. During the cooler seasons, they allow finely regulated cross-ventilation. Two fireplaces in the Esherick House provide both "night light" and warmth: one at the south wall of the living room, and the other in an upstairs bathroom. There, you can lie in the bathtub and enjoy a real chimney fire (analytical sketches no. 18 and no. 31). By means of a large wooden drawer which moves horizontally into the wall, this bathtub converts into a sofa. The house is heated by a sophisticated hot-air system which, however, does not function as air conditioning in summer.

Louis I. Kahn designed two further projects for the Esherick Family: the unbuilt addition to the existing M. Esherick House, and the atelier constructed for the artist Wharton Esherick.

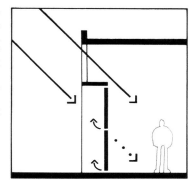

a. 北立面
b. 东南立面
c1./c2. 分析草图
d. 底层平面图；顶层平面图；剖面图
e. 西北立面
f. 计划加建设计图
g1./g2. 起居室（通风嵌板打开）

a. **North facade**
b. **Southeast facade**
c1./c2. **Analytical drawings**
d. **Ground floor; top floor; section**
e. **Northwest facade**
f. **Planned addition**
g1./g2. **Living room (with ventilation panel open)**

g1.

g2.

第一唯一神教堂及学校
1959–1969；建成
罗切斯特，纽约州

FIRST UNITARIAN
CHURCH AND SCHOOL
1959 – 1969; BUILT
ROCHESTER, NEW YORK

第一唯一神教堂及学校分两个阶段建造。第一阶段建造了西侧的大礼拜堂、外围的教室、接待室、厨房和图书馆。东侧的建造为第二阶段。两个阶段在立面和平面上都有显著的区别。康或许在内心决定要参照他1959年参观的位于法国南部的阿尔比大教堂，因此设计了阶梯形的砖墙。

大礼拜堂的平面基于一个几乎完美的正方形。屋顶结构以教室内墙支撑。礼拜堂6米高的混凝土砌块墙外面是环绕着走廊布置的教室。四个大天窗使阳光从四角射入礼拜堂，不然这个房间就没有什么窗户了。从而，这个房间的氛围随着太阳高度、天气和季节发生着变化。康第一次成功地使一间无窗的房间仅依赖顶部的自然采光。"很哥特，是不是？它令你不安吗？我个人很喜欢。"（16）

西侧的立面进行了明显的三维调整，这是对莫里斯住宅的一个符合逻辑的改进。窗口用陶瓦封闭起来：因此窗玻璃装在墙的内侧。在康的草图中，窗口的基脚区域和一楼以上后退的墙体以某个倾斜的角度分割开来。然而在实际建造中，这些角度变成了直角。东侧在几年后建成，也以窗口为主要特征，尽管它

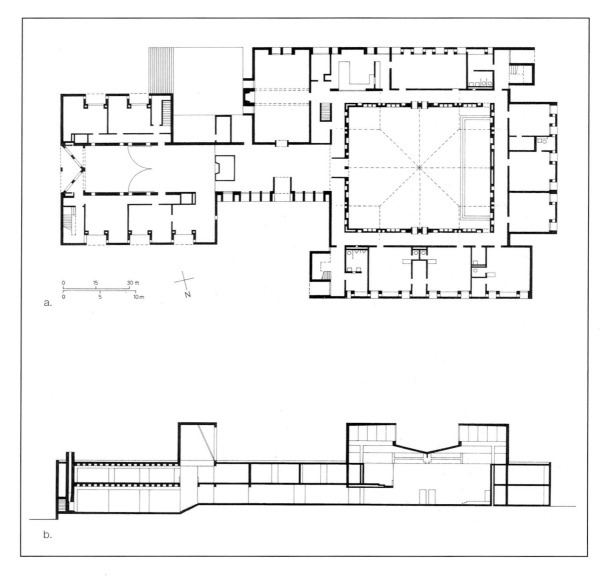

a.

b.

The First Unitarian Church and School was built in two stages. The first phase implemented the west wing with large sanctuary, classrooms in a peripheral configuration, reception room, kitchen, and library. Construction of the east wing followed as second stage. The differences between the two stages are clearly evidenced in the facades and floor plans. Kahn perhaps decided for the inwardly stepped brick wall on the basis of his visit to the Cathedral of Albi in southern France in 1959.

The plan of the large sanctuary is based on an almost perfectly square form. The roof structurally rests on the inside walls of the classrooms. The circumferential corridor for the classrooms is accommodated behind a 6-meter-high wall of concrete blocks. Four great skylights in the corners of the sanctuary admit light into the otherwise windowless room. The atmosphere in its interior therefore changes according to the height of the sun, the weather, and the seasons. For the first time, Kahn achieved success with a windowless interior room illuminated only by natural light falling from above. "It's very Gothic, isn't it? Does that bother you? I like it myself." (16)

The facade of the west wing with its pronounced three-dimensional modulation is a further logical development of the Morris House. The window niches are closed off above by terra-cotta tiles: the window glass is accordingly installed at the inside plane of the walls. In schematic drawings by Kahn, the footing areas in the

c.

d.

a. 平面图
b. 剖面图
c. 西南立面
d. 西立面

a. Plan
b. Section
c. Southwest facade
d. West facade

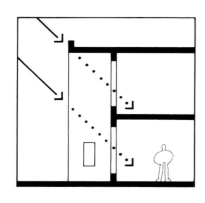

们没有西侧的窗口设计得精致；相反，它们被重新诠释为巨大的"采光口"。东立面上有一个壁龛，其中的窗户呈"V"形布置。此处的一个重要原则是利用墙上的壁龛为窗口遮挡直射阳光和风雨。从而使窗户成为了建筑中一个用来表达的要素，而不仅仅是墙上的一个洞口。康写道：

"我们再次感受到阳光的炙热，学会每时每刻感受耀眼的眩光……无论阳光是照耀在罗切斯特还是罗安达，都是一种领悟……如果你望着一幢文艺复兴式建筑……它的窗户在建筑中格外突出……一个这种形式的窗户……一个嵌入洞口中的窗户……这非常棒，因为它让光线能从两边射入，让阳光不那么刺眼。"（17）
另外值得注意的是首层壁龛两侧的小窗户。它们能让人安全地坐在壁龛里向外"眺望"。

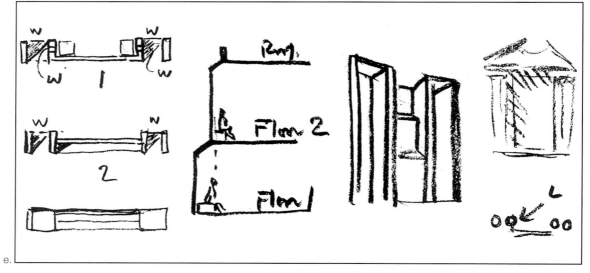

e. 说明性草图
f. 底层窗户细部
g. 窗口
h. 内部
i. 采光塔

e. Explanatory sketches
f. Detail of window on ground floor
g. Window niches
h. Interior
i. Light tower

window niches and the wall setbacks above the ground floor have been set off at an oblique angle. As actually constructed, however, the angles are straight. The east wing built years later also features window niches, although they are less delicately designed than those in the west wing; instead, they have been reinterpreted as giant "light ports." The east facade has one niche in which the window is hung in a V-arrangement. An essential principle here is the concept of using wall niches for protection of windows from direct sunlight and the elements. The window consequently becomes an architecturally articulated element, rather than merely an opening in the wall. Kahn writes:
"We felt the starkness of light again, learning also to be conscious of glare every time ... whether it's the glare in Rochester or glare in Luanda, it still was one realization.... If you looked at a Renaissance building ... in which a window has been highly accentuated architecturally ... a window that's made in this form... windows framed into the opening.... This was very good because it allowed the light that came in on the sides to help again to modify the glare." (17)

Also noteworthy are the small windows in the sides of the niches on the ground floor. They allow one to sit protected in the niches and "to keep a lookout" toward the outside.

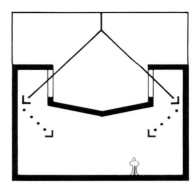

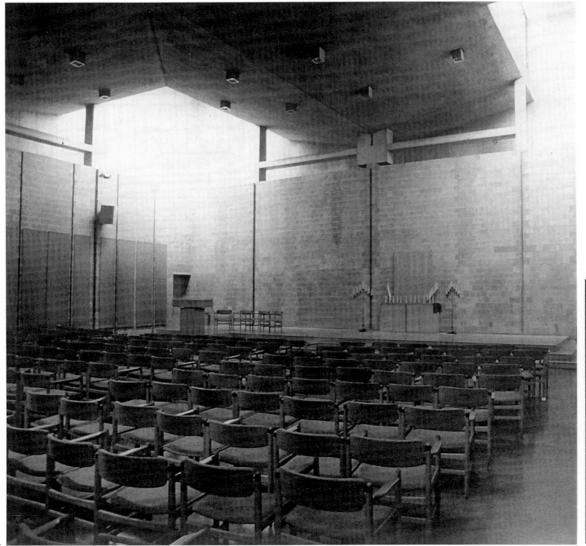

h.

i.

美国领事馆及馆舍住宅
1959–1962；未建成
罗安达，安哥拉

U.S. CONSULATE AND
RESIDENCE
1959 – 1962; UNBUILT
LUANDA, ANGOLA

"我正在非洲离赤道非常近的地方设计一幢建筑。那里刺目的阳光颇具危险性：每个人在阳光下看起来都是黑色的。我们需要阳光，但它也是一种危害。"（18）

位于同一块地上的这两幢建筑的主要特征是玻璃窗前面用作遮阳板的外墙以及新发明的遮阳屋顶。用地南侧的办公楼，在形式和结构上都比北侧的馆舍住宅部分更加稳重。

东、西立面每面各有四片独立的混凝土板（参阅弗莱舍住宅）。北侧和南侧是没有窗户的实墙，这些墙面被深深凹进去的入口分割开来。支撑着遮阳屋顶的八根柱子完全与建筑本身相互独立：这样做的优点在于没有柱子穿透办公室部分的屋顶。直至今日，这个项目仍代表了康最符合逻辑的设计，他成功找到了许多解决问题的方法，例如针对直射阳光、室内外的视线联系以及怎样遮风避雨。尽管路易斯·I.康后来在相似的气候条件下实际建成了一些建筑，奇怪的是他再也没有采用这个遮阳屋顶的理念。

"我觉得应该将挡雨的屋顶与遮阳的屋顶分开，我是在告诉街上的人们他们的生活方式。"（19）

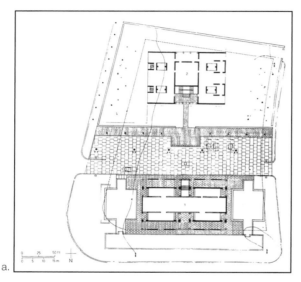
a.

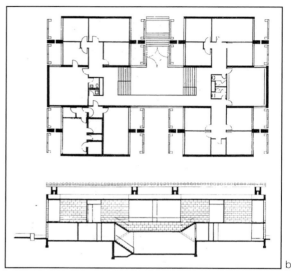
b.

c.

a. 总平面图
b. 平面图；剖面图
c. 模型
d. 轴测图
e. 说明性草图

a. Site plan
b. Plan; section
c. Model
d. Axonometric projection
e. Explanatory sketch

"I am doing a building in Africa which is very close to the equator. The glare is killing: everybody looks black against the sunlight. Light is a needed thing, but still an enemy." (18)

Distinctive for these two buildings on one plot are the brisesoleil walls in front of the glazed areas, and the newly developed sunshade roof. The administration building on the south side of the plot is formally and structurally more logically consistent than is the residential wing to the north.

The east and west facades are each protected by four freestanding concrete slabs (cf. the Fisher House). The north and south sides consist of windowless walls divided by the deep setback of the entrance. The eight pillars which support the sunshade roof are completely independent of the building itself: with the advantage that no pillars penetrate the roof surface of the office wing. Until now, this project represents the most logically consistent of Kahn's designs with respect to successfully thinking through solutions for problems such as directly incident sunlight, visual relationships between inside and outside, and protection from the elements. Although Louis I. Kahn later actually executed buildings under similar climatic conditions, he strangely never resorted again to this concept of a sunshade roof.

"I feel that in bringing the rain roof and the sun roof away from each other, I was telling the man on the street his way of life." (19)

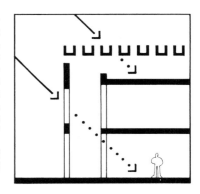

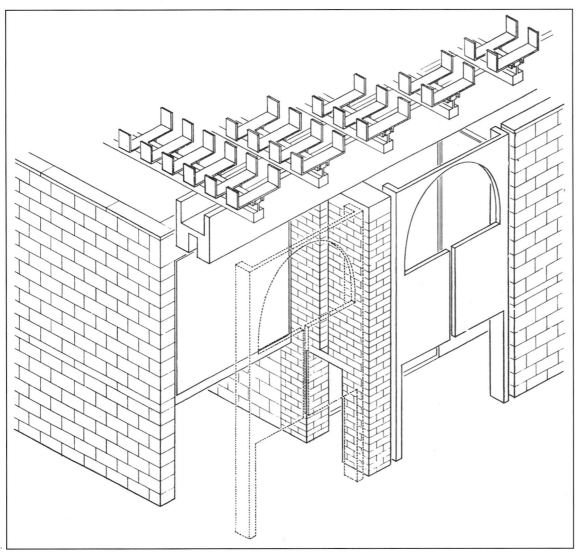

d.

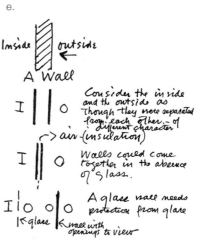

e.

索克生物研究所
1959–1965；
实验室部分建成
拉霍亚，加利福尼亚州

SALK INSTITUTE FOR
BIOLOGICAL STUDIES
1959 – 1965;
LABORATORY BUILT
LA JOLLA, CALIFORNIA

研究所用地位于一处悬崖旁边，陡峭的悬崖坠入太平洋。最初设计中有三组建筑，最终只有实验室部分建成了：会议中心和科学家住宅最终都没有落成。两幢矩形实验楼的布置在它们与大海之间形成了开阔的空间：一个很难与辉煌联系在一起的地方。康的研究所设计最初是基于塔楼的理念（参阅理查德医学研究中心）。然而，从费城塔楼建筑中获取的在功能方面的设计经验以及所遇到的照明问题，使康有了新的领悟，并在拉霍亚找到了一种全新的解决方案。为了实现中间的预留空间，康再一次采用了空腹桁架技术：在此处梁中空腹形成的空间尺度较大，让使用者完全能够毫无困难地直立行走。最终得到一个无柱的空间，成功地满足了空间的灵活性需求。康把梁间的间隙空间利用起来，用作服务用房；悬臂梁的采用为低处实验室的玻璃墙面遮挡了直射阳光。在科学家们面向大海的研究用房和图书馆里，康用一种新的方式安装了水平滑动木质嵌板（参阅太阳屋设计）。这些遮阳构件可以根据使用者的需要进行调节。整个窗扇都可以滑动，以便获得通风；也可以使用防蚊纱窗。

a. 平面图；剖面图
b. 庭院
c. 模型照片
d. 西南立面

a. **Plan; section**
b. **Court**
c. **Picture of the model**
d. **Southwest facade**

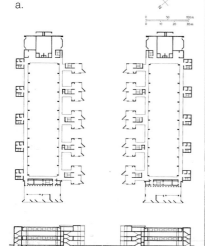

The plot of the research institute lies on the edge of a cliff which descends steeply to the Pacific. Of the originally planned three complexes, only the laboratories were built: the conference center and the residence for scientists were never executed. The two oblong buildings for the laboratories are arranged such that an open space is formed between them and the ocean: a space hardly equalled for splendor anywhere. Kahn's design for the research institute was originally based on the tower principle (cf. the Richards Medical Research Laboratories). Experience gained with the functional and lighting problems encountered with the tower constructions in Philadelphia, however, led Kahn to new insights and to a completely new solution in La Jolla.

Kahn once again employed Vierendeel techniques here, for a special room reserved for the media: but in this case, with such generous dimensioning that the occupants in this room have no difficulty in walking fully upright. The result is a room, without columns, which successfully ensures the required flexibility. Kahn provides for finishing of the unoccupied rooms resulting here to furnish floor space for service functions; the cantilever solution implemented for this space protects the glass walls of the lower-lying laboratory rooms from direct sunlight. In the scientists' study cubicles situated toward the ocean, and in the library rooms, Kahn installed horizontally sliding wooden panels in a novel manner (cf. the design for the Solar House). These sun-

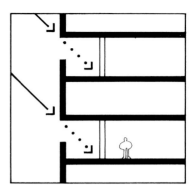

d.

打算建在场地西北侧的会场像一个修道院。这组建筑包括礼堂、客房、厨房和餐厅、会议室、图书馆以及一个散步的花园。在图书馆和餐厅前独立休息室的玻璃幕墙前，采用了一层"遮阳墙"来遮挡直射阳光（参阅美国领事馆及馆舍住宅）。在一张草图上，康详细阐述了这种为存在眩光问题的房间采取的建筑解决方案。混凝土墙体上切出的半圆形洞口将刺目的直射阳光打散，让人能眺望外面，并投下美丽的阴影。

"我开始意识到，每一个窗户都应当面对一面独立的墙。白天的阳光照射在这面墙上，它能够大胆地向天空敞开。这面墙削弱了眩光，却没有阻断视野。这样，就避免了靠近窗户的天窗格栅投下的分散的光斑所形成的强烈对比。"（20）

e. 平面图
f. 透视图（晚期版本）
g. 模型
h. 说明性草图
i. 图书馆（早期版本）
j1./j2. 图书馆；模型照片

e. Plan
f. Perspective (late version)
g. Model
h. Explanatory sketch
i. Library (early version)
j1./j2. Library; pictures of the model

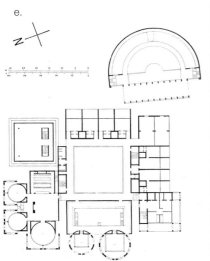

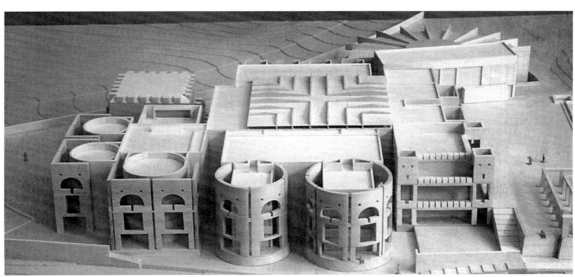

shade elements feature panels which can be adjusted as the occupant wishes. To provide ventilation, one can horizontally slide the entire window frames; the same is possible for the mosquito screening.
The design of the Meeting Place, intended for the northwest of the plot, resembles a cloister. This complex includes auditorium, guest accommodations, kitchen with dining rooms, seminar rooms, library, and a garden for strolling. The freestanding, fully glazed day rooms in front of the library and the dining rooms are protected from sunlight by a shell of "sun walls" (cf. U.S. Consulate and Residence). On a sketch, Kahn elaborated on this architectural solution for rooms confronted with the problem of glare. The semicircular openings cut out of the concrete shells diffract the directly incident light, allow a view of the outside, and cast ornamental shadows.
"I came to the realization that every window should have a free wall to face. This wall, receiving the light of day, would have a bold opening to the sky. The glare is modified by the lighted wall, and the view is not shut off. In this way, the contrast made by separated patterns of glare, which skylight grilles close to the window make, is avoided." (20)

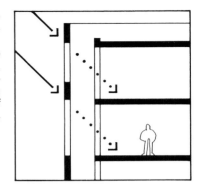

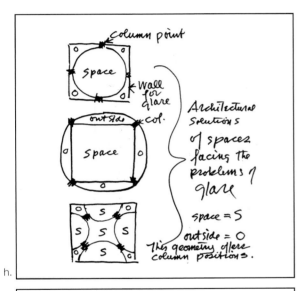

h.

j1.

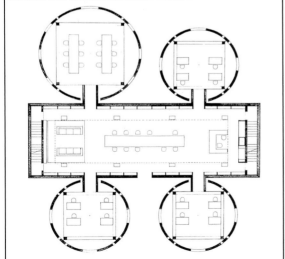

i.

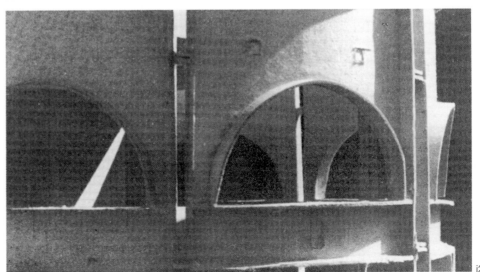

j2.

表演艺术中心
1959–1973；建成
韦恩堡市，
印第安纳州

PERFORMING ARTS
CENTER
1959 – 1973; BUILT
FORT WAYNE, INDIANA

"唯有自然光铸就了建筑。"（21）然而，路易斯·I. 康也擅长设计电影院、报告厅以及其他礼堂建筑等合理采用人工照明的建筑。仅靠工程技术就能使光线影响这些空间的特质，而非季节、太阳高度或是飘过的云朵。此处，康没有寻求和谐地使用夜晚的自然光——火光来照明，而是用同样透彻的方式，引入了与传统技术手段对应的现代技术性的光线。康在处理人工光方面表现出极大的自主性。（参阅韦斯住宅、耶鲁大学美术馆、厄德曼礼堂以及耶鲁英国艺术中心）。然而，对稳定的人工照明的控制很难与多变的自然光相比较。这个项目仅仅是代表了康在人工照明方面的能力，也是他的工作中值得探讨的一个方面。

入口立面或许可以诠释为一个超大尺度的古典戏剧面具符号（眼睛、鼻子和嘴）。在剧场演出前后进进出出观众的影子被投在外面被照亮的橱窗上（参阅康设计的拉布·杜尔电影院，本书没有收录）。在这个初始的外向表皮后面，是独立的剧院观众厅和舞台，在声学上和结构上都与外壳相分离："盒子里的盒子"。屋顶部分的空腔中收纳了聚光灯和通风系

a.

b.

"Natural light is the only light that makes architecture." (21) Nevertheless, Louis I. Kahn also excelled in the design of legitimate and motion-picture theaters, lecture halls, and other auditoriums, i.e., rooms with artificial illumination. Engineering alone can influence the character lent by light to such rooms and not the seasons, the height of the sun, or passing clouds. Here it is not fire as the "light of the night" which Kahn seeks to harmonize: rather, he addresses its modern counterpart, technical light, with the same thoroughness. Kahn demonstrated great sovereignty in handling of artificial light (cf. Weiss House, Yale University Art Gallery, Erdman Hall, and the Yale Center for British Art). Control of static artificial illumination is hardly comparable, however, to that of dynamic natural light. This project is merely representative of Kahn's expertise with artificial illumination, an aspect of his work worthy of treatment by itself.

The entrance facade may be interpreted as an overdimensioned symbol for the antique dramatic mask (eyes, nose, and mouth). Before and after evening theater performances, the coming and going of theatergoers is projected toward the outside as in an illuminated show window (cf. Kahn's design for the Raab Dual Movie Theater, not documented here). Behind this initial, extroverted envelope is the autonomous theater auditorium and stage, detached both acoustically and structurally from the outer shell: "the box in the box." The hollow

a. 南立面
b. 前厅
c. 前厅细部

a. South facade
b. Foyer
c. Detail foyer

统,通过狭窄的走廊进入。狭窄走廊的概念令人想起理查德医学研究中心以及索克生物研究所夹层式的中央空间。

观众厅的结构像一把小提琴:外部的砖壳与观众厅相分离,像是琴盒与里面的提琴。斜角满足了多种需求:结构需求、功能需求——在其中设置了灯光桥架以及(尤其是)声学上的需求。

值得注意的是,路易斯·I.康在这个项目上工作了14年,期间由于资金限制大大缩小了项目规模。

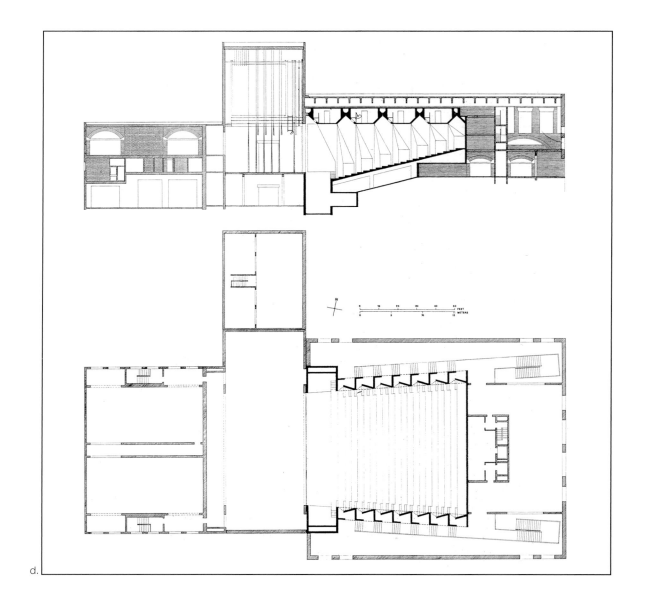

d. 平面图和剖面图
e. 剧院观众厅

d. Plan and section
e. Theater auditorium

space in the ceiling accommodates the spotlights and the ventilation system, accessed via catwalks. The concept for the catwalks calls to mind the media rooms in the Richards Medical Research Building and the Salk Institute.

The auditorium is structured like a violin: the exterior brick shell is separated from the auditorium, like the carrying case from the violin inside. The oblique angles fulfill multiple purposes: a structural, a functional in that they form light bridges, and (not least) an acoustic function.

It is noteworthy that Louis I. Kahn worked on this project for 14 years, during which financial constraints extensively reduced its scope.

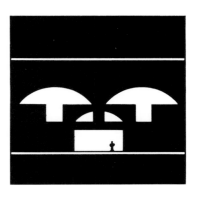

e.

埃莉诺·唐纳利·厄德曼礼堂
1960–1965；建成
布林·茅尔学院，
宾夕法尼亚州

ELEANOR DONNELLY
ERDMAN HALL
1960 – 1965; BUILT
BRYN MAWR COLLEGE;
PENNSYLVANIA

20世纪60年代，康经常提及邓弗里斯郡的苏格兰城堡康隆根，以阐明他对于中央公共区域的理念，将小房间明显地布置四周。在布林·茅尔，康的设计与第一唯一神教堂一样，是基于一个环绕着小房间的中心庭院。十二座巨大的塔楼使阳光进入三个正方形建筑的中央空间，这三个正方形建筑在两个角上相互重叠。与罗切斯特的教堂不同，天窗玻璃朝向外面，使得从底层可以直接看到天空。由于光线在一天中任何时候都不会发生衍射，就不会产生过度的神圣氛围。

内嵌的立面令人想起堡垒。一种城堞似的结构封闭了顶层垂直的窗户。在内嵌立面以上的房间，窗户洞口是横向设置的。当你进入一个这样的房间中，起初眺望窗外的视野被中轴线上的墙挡住了；然而，左右两侧的采光槽，给人留下了更深刻的印象。建筑中靠内部的房间利用中间插入的窗户采光。织物卷帘装在窗户内侧，遮挡了直射阳光。令人惊讶的是，窗子只能靠曲柄向外打开；当窗户打开时，整个房间就暴露在风雨中了。

a. 平面图
b. 剖面图
c. 屋顶顶视图
d. 前厅
e. 采光塔

a. Plan
b. Section
c. Top view of the roof
d. Foyer
e. Light tower

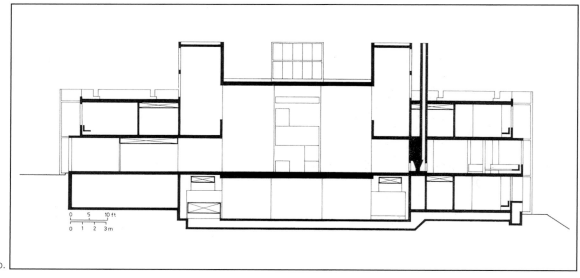

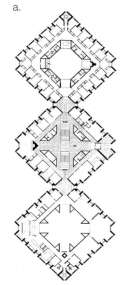

During the 1960's Kahn often referred to the Scottish castle Comlogan in Dumphriesshire to clarify his concept of a central common area with a pronounced periphery of smaller rooms. At Bryn Mawr, Kahn's design, as with the First Unitarian Church, is strongly based on a central courtyard surrounded by rooms. Twelve large towers admit sunlight into the central space of the three square constructions, which intrude into each other at two corners. Unlike the church in Rochester, the skylight glazing is oriented toward the outside a configuration which allows, from the ground floor, a direct view of the sky. Since light is not diffracted in the light shafts at all times of day, an excessive sacred atmosphere cannot prevail.

The inset facade creates the impression of a citadel. A kind of merlon closes off the vertically oriented top-floor windows. In the rooms extending beyond the inset facade, the window openings are installed laterally. When one enters one of these rooms, the view out is initially blocked by the wall installed on the central axis; the light niches to the left and right, however, make a correspondingly greater impression. The rooms one layer deeper in the building are illuminated by inset central windows. Rolling fabric shades installed inside the windows furnish protection from the sun. Surprisingly, the windows open only toward the outside, by a crank; when opened, they are exposed to the elements.

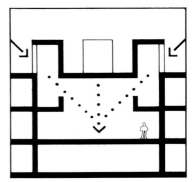

d.

e.

费城艺术学院
1960–1966；未建成
费城，
宾夕法尼亚州

PHILADELPHIA
COLLEGE OF ART
1960 – 1966; UNBUILT
PHILADELPHIA,
PENNSYLVANIA

a. 平面图（晚期版本）
b. 模型顶视图（晚期版本）
c. 模型剖面图（早期版本）
d. 说明性草图

a. Plan (late version)
b. Top view of the model (late version)
c. Section of the model (early version)
d. Explanatory sketch

费城艺术学院计划扩建其位于市中心布罗德大街的设施。学院打算保留现有的三幢建筑，这三幢建筑是在19世纪由建筑师弗兰克·海林·弗内斯（Frank Heyling Furness，1839—1912，美国建筑师）、乔治·沃森·休伊特（George Wattson Hewitt，1841-1916，美国建筑师）和约翰·哈维兰（John Haviland，1792—1852，美国建筑师）设计建造的。复杂而大量的空间需求，迫使康必须设计高层建筑，并与老建筑共存。由于剩下的开放空间太小，康设计了特别的天井，或者他将其称为"内庭院"。

这里的平面图和模型照片并不完全相同。楼层平面图是最后的版本，其中的大中央庭院是矩形的。而展示的模型是早期版本，显示了康让阳光向下射入建筑中心的理念。康设计了巨大的采光井为南向教室和走廊提供自然光，另外，让阳光经过半圆柱上的三角形切口从后部进入工作室。画家、雕塑家和建筑师的工作室朝北，由此获得理想的中性而稳定的光线。部分玻璃倾斜安装，成功地满足了艺术家对光线品质的需求。康大量强调了走在建筑中时的视线联系，并且能在工作室看到正在工作的学生。

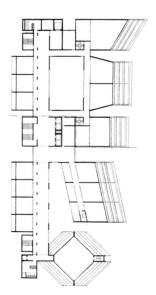

The Philadelphia College of Art planned to expand its midtown facilities, located on a block of Broad Street. The College intended to retain three existing buildings there, built by the architects Furness, Hewitt, and Haviland in the nineteenth century. Complex and extensive space requirements forced Louis I. Kahn to plan highrise buildings to coexist with the old construction. The remaining open space was so small that Kahn designed special patios, or what he called "inside gardens."

The plans and the model pictured here are not fully identical. The floor plans represent the last version, in which the large central courtyard is rectangular. The models shown here are early versions which illustrate Kahn's concept of admitting light down into the center of the buildings. The huge light shafts which Kahn planned provide natural light to the classrooms and corridors oriented to the south, and additionally admit light through triangles cut out of half cylinders from the rear into the ateliers. Studios for the painters, sculptors, architects, and other artists optimally receive neutral, uniform light owing to their orientation to the north. The glass surfaces, some of them installed at an angle, successfully satisfy the expectations held by such artists for light characteristics. Kahn placed great emphasis on the visual contacts provided while walking through the buildings, and on the ability to see the students at work in their studios.

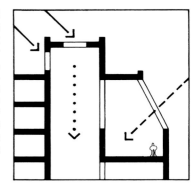

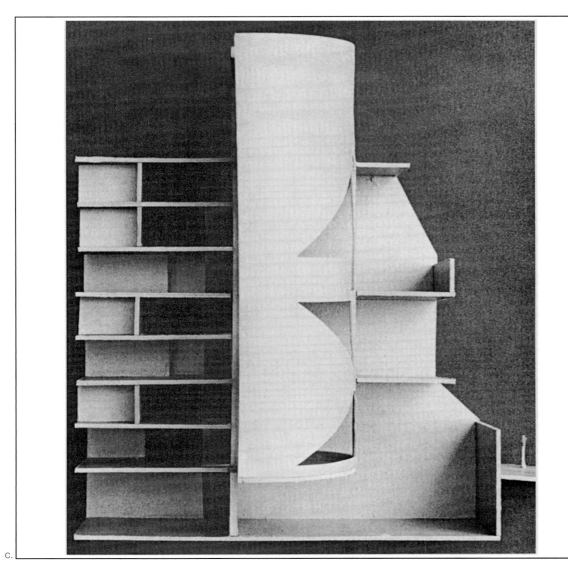

c.

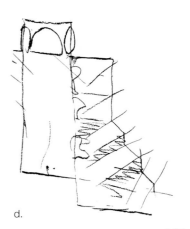

d.

诺曼·费舍住宅
1960–1967；建成
哈特伯勒，
宾夕法尼亚州

NORMAN FISHER HOUSE
1960 – 1967; BUILT
HATBORO,
PENNSYLVANIA

a. 底层平面图；顶层平面图
b. 西南立面
c. 起居室
d. 窗户和通风嵌板细部

a. Ground floor; top floor
b. Southwest facade
c. Living room
d. Detail of window and ventilation panel

路易斯·I. 康用了六年时间设计这幢住宅，从许多角度来看，其结果都令人惊讶。同以前一样，康在这个设计中使几何形相互之间呈独特的偏转角度。他还遵循了一个设计原则，即将两层高的公共起居室和餐厅区域与私密的卧室空间清晰地分开，其中一个主要原因是起居室和餐厅的位置与一条小溪平行，而卧室正好是东西向的。此处，转角窗（参阅奥泽住宅和罗奇住宅）与木质通风嵌板以及固定的长凳，几乎成为一个雕塑般的整体。这个窗户两层多高，直至天花板，能够很好地看到公园般的庭院。由此转角窗成为了一扇全景窗。除了三个垂直的照明和通风槽外，面向街道的立面是封闭的。落水管在水平方向上稍微错开，几乎以一种蒙德里安式构图布置在立面上。不能开启的窗户和独立的通风嵌板安装在壁龛内（参阅埃西里科住宅）。沿着东立面，在楼上的两间卧室里不开启的窗户、通风嵌板、桌子以及衣橱和谐地组织在一起，形成了一件绝妙的、多功能的"家具"。墙体作为室内与室外之间三维的过滤器，已经被理想地实现了。

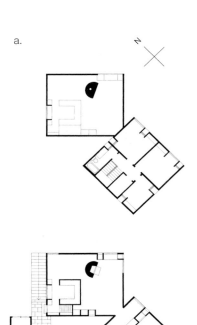

a.

b.

Louis I. Kahn spent six years on the design for this home, with surprising results from a number of viewpoints. As before, Kahn works here with geometric forms rotated at unusual angles to each other. He also follows the principle of distinct separation of the communal two-story living and dining areas from the private bedroom space, one primary reason being location of the living and dining room parallel to a brook, and bedroom orientation exactly in an east-west direction. Here, the corner window (also cf. Oser House and Roche House), together with the wooden ventilation panels and built-in bench, become almost a sculptural entity. This window extends over two stories, up to the ceiling, and provides a wonderful view onto the park-like yard. The corner window thereby becomes a panorama window. With the exception of three vertical light and ventilation slots, the facades toward the street are closed. The water-drip molds, horizontally slightly staggered, arrange the facade almost in the manner of a Mondrian picture. The non-opening windows and the separate ventilation panels are installed in the niches (cf. Esherick House). Along the east facade, the non-opening windows, the ventilation panels, the table, and the wardrobe chest in the two upper-story bedrooms are harmoniously integrated to form one great, multifunctional piece of "furniture." The wall as three-dimensional filter between indoors and out has been ideally realized.

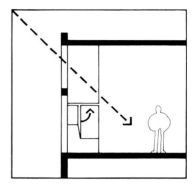

c.

d.

米克韦以色列犹太教会堂
1961–1972；未建成
费城，
宾夕法尼亚州

**MIKVEH ISRAEL
SYNAGOGUE
1961 – 1972; UNBUILT
PHILADELPHIA,
PENNSYLVANIA**

路易斯·I. 康为这座犹太教会堂的设计工作了十一年多。这个项目要在费城的一个历史街区中实现它复杂的空间需求，其中包括犹太教堂（本文所述）、礼拜堂、苏克棚、礼堂、博物馆、教室、办公室以及其他设施。尽管康的总体设计与修改返工了大约十次，但圆形"光塔"的理念（22）贯穿始终。此处展示的平面图是1964—1966年间的第七版，它是基于一个拉长了的八边形，其中嵌入一个椭圆形。尽管康最初的草图令人想起阿尔比大教堂，但这座犹太教会堂的窗户是在塔柱处，而不是在平直的部分。

带有整体托梁的凸透镜形状的屋顶结构横跨在椭圆形会堂上，以封闭的平直墙体支撑。这种看起来像是天花板悬挂在房间上空的手法，是康的一种新的建筑形式语汇。室内透视将天花板描绘成光滑的白色表面。然而如果实际建成，墙上的洞口会在这个表面上投下明显的光影图案。礼拜堂天花板给人留下的印象或许可以相当于横跨在公共场所上方的天篷。在胡瓦犹太教会堂的设计中，康进一步发展了曲面天花板的理念，使更多的光线进入内部空间。康称之为"光瓶"或"窗室"的部分，也充当了整体空间

a.

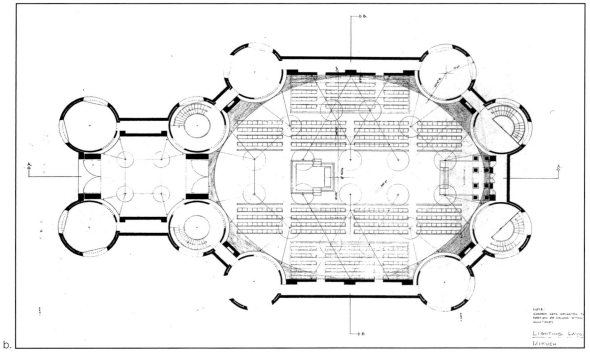

b.

Louis I. Kahn labored on the design for this synagogue for over eleven long years. Complex space requirements were planned for realization in an historical part of Philadelphia, to include the synagogue (documented here), a chapel, a sukkah, an auditorium, a museum, classrooms, offices, and other facilities. Although Kahn's overall design was modified and reworked about ten times, the concept of the round "light towers" (22) prevailed throughout. The floor plan shown here is from the seventh version, from 1964 - 1966, and is based on an elongated octagon in which an ellipse has been inscribed. Although Kahn's first sketches call to mind the Cathedral of Albi, the windows in this synagogue are in the tower cylinders, and not in the straight sections of wall.

A lens-formed ceiling structure with integrated joists spans the elliptical synagogue chamber and rests on the closed, straight wall slabs. This view of the underside of a ceiling, suspended over a room, is new in Kahn's vocabulary of architectural forms. The interior perspective depicts the underside of the ceiling as a smooth, white surface. If actually constructed, however, the wall openings would cast starkly pronounced patterns of shadow and light onto this surface. The impression created by the sanctuary ceiling is perhaps comparable to that of a canopy spanned over a public place. In his design for the Hurva Synagogue, Kahn further developed the concept of a sweeping ceiling to enhance admission of light into the middle of the room.

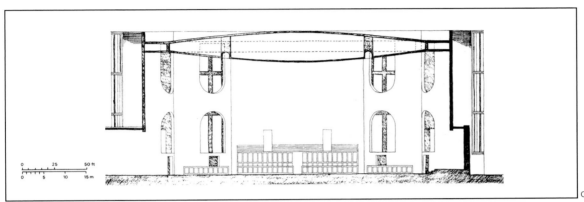
c.

d.

a. 立面图
b. 平面图
c. 剖面图
d. 内部

a. Elevation
b. Plan
c. Section
d. Interior

中的一处连接或节点，而不仅仅是为内部提供采光。巨大的光塔外侧设有窗户，窗口是一道砖拱。每个圆柱体顶部都用混凝土板封闭。圆柱塔内侧没有玻璃的洞口让光线像经过了一个巨大的过滤器再进入室内。因此墙体表面在光线映衬下显得很幽暗，使内部洞口形状看起来被放大了。光塔细部图中左侧展示的是窗格；右侧描述的是与圆柱体相分离的玻璃表面。

"这些空间被直径 20 英尺的窗室包围，窗室之间通过有墙的走廊相连。这些窗室部分外侧的洞口装有玻璃，面向内部的则是巨大的没有玻璃的拱形洞口。这些光的房间环绕着礼拜堂，它们还被用作女性通行的高大空间。这些窗室构建了入口门厅和对面的礼拜堂……窗户外侧并不承重；你能在平面图中看到，窗户之间的部分承重。"

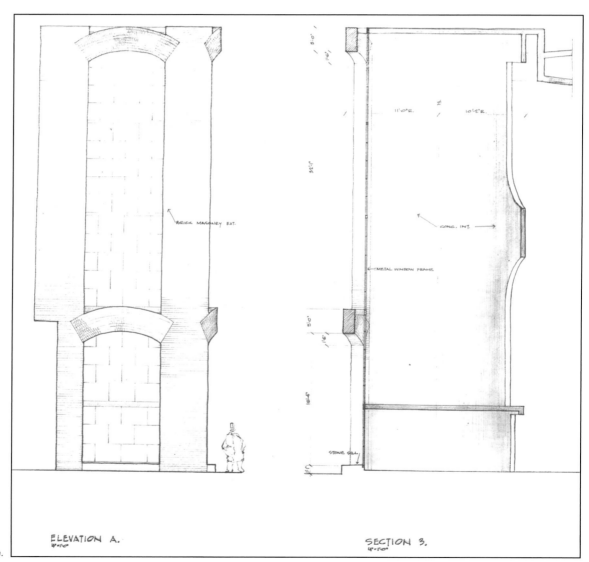

e.

What Kahn called the "light bottles" or "window rooms" additionally serve as a kind of joint or hinge for the entire space, and not only for illumination of the interior. The mighty light towers laterally accommodate the windows, which are framed by brick arches. Each cylinder is closed at the top by a concrete cover. The nonglazed openings at the inside of the towers admit light into the interior as through a giant filter. The wall surfaces therefore stand off dark against the light, with resulting amplification of the forms of the inner openings. The detail drawing of a light tower shows in the left view the window muntins; the section on the right depicts the glass surface detached from the cylinder form.

"The spaces are enclosed by window rooms twenty feet in diameter connected by walled passages. These window room elements have glazed openings on one exterior side and larger unglazed arched openings facing the interior. These rooms of light surrounding the synagogue chamber serve as an ambulatory and the high places for women. These window rooms prevail in the composition of the entrance chamber and the chapel across the way.... The windows on the outside do not support the building; what supports the building, as you can see on the plan, are the spaces between the windows."

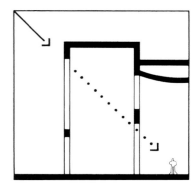

e. 圆柱形光塔
f. 模型内部照片

e. **Light cylinder**
f. **Picture of model interior**

印度管理学院
1962–1974；建成
艾哈迈达巴德，印度

INDIAN INSTITUTE OF
MANAGEMENT
1962 – 1974; BUILT
AHMEDABAD, INDIA

起初，巴克里希纳·多西（Balkrishna Doshi, 1927—，印度建筑师）受邀设计建造印度管理学院。然而当时已在艾哈迈达巴德与勒·柯布西耶合作的多西推荐了路易斯·I. 康作为项目的建筑师。这项设计最初是以艾哈迈达巴德设计学院的工作室任务的形式，作为合作研究项目，并进一步由费城路易斯·I. 康事务所的印度学生付诸实践。

最初设计是在学校周边的一块平地上建造一组建筑，包括教室、办公室、厨房和餐厅、图书馆、宿舍、教师公寓以及员工宿舍。还设计了一个人工湖将教室和宿舍区域与建筑群的其余部分隔开，但最终没有建成。

"太阳的热量、风、光、雨、尘"：康将这五个关键词写在1962年11月14日的一张草图上。（23）这些严苛的气候条件极大地影响了多年来艾哈迈达巴德的规划和建筑设计。

我们可以将两个单体建筑当作整个复杂建筑群中的典例：图书馆和一个宿舍。值得注意的是，大部分建筑都是东西向的，以顺应主导风向。U 形的建筑综合体中包括教室、图书馆和管理设施，这幢建筑

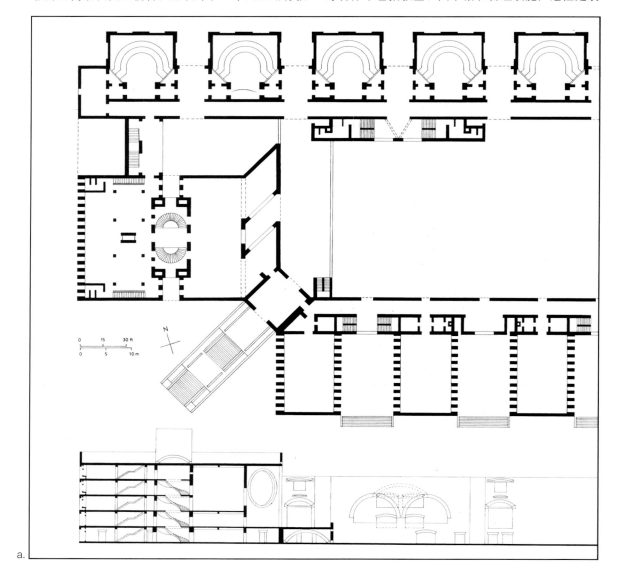

a. 平面图和剖面图
b. 庭院
c. 说明性草图（晚期版本）

a. Plan and section
b. Court
c. Explanatory sketch (late version)

a.

Originally, B. Doshi was invited to design and build the Indian Institute of Management. Doshi, however, who had already collaborated with Le Corbusier in Ahmedabad, recommended Louis I. Kahn as architect. The design was prepared as a joint project, in the form of a studio assignment at the Institute of Design in Ahmedabad, and was further developed to the execution stage by Indian students in Kahn's office in Philadelphia.

Original planning was to build a complex on a flat area at the periphery of the campus, to include classrooms, offices, a dining hall with kitchen, a library, dormitories, apartments for lecturers, and staff accommodations. An artificial lake, designed to separate the classroom and dormitory area from the rest of the complex, was never executed.

"Sun-heat, wind, light, rain, dust": Kahn noted these five key words on 14 November 1962 on one of his first sketches. (23) These severe climatic constraints significantly influenced the Ahmedabad design over several years of planning and construction.

We may consider two individual buildings as examples for the entire intricate complex: the library and one of the dormitories. It is noteworthy that most of the buildings are oriented east-west, to benefit from the prevailing wind direction. The U-shaped building complex with classrooms, the library, and the administration facilities, was originally designed to be closed, like a cloister, around a courtyard. The dining room and the kitchen were later built on the southeast

b.

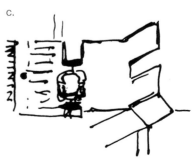

c.

最初的设计是封闭的，像一圈环绕着庭院的回廊。餐厅和厨房后来建在场地的东南侧。如今，入口平台以及容纳着图书馆"通风采光口"的巨大幕墙，封闭了广场。楼层平面图清晰地展示了墙体如何从正交几何形中旋转出来。这些墙面向风的方向（由西向东），并通过三个圆形窗洞来调节射入图书馆三层阅览室的阳光。这幢建筑呈现出了明显的较深的空间区域及双层墙体的特征。呈一定角度设置的翼墙实际上起到遮阳的作用：此处它们不是独立的元素，而是整个空间的组成部分。康在这幢建筑中，以一种极其复杂的方式解决了炫目阳光的衍射问题。圆形的洞口像是从墙上打出来的一样，成为了建筑中一个尺寸巨大的构件。砖砌的圆环由预制混凝土托梁固定。康将传统的砖与现代的混凝土融合在一起，形成一个和谐的整体。图书馆圆形的窗洞原本是开敞的，但后来装上了玻璃。由于强劲的季风和来来去去的鸽子，这种措施是十分必要的，尽管这使得内部空间失去了古朴的特质。

宿舍设计中，两翼的侧墙形成了一个漏斗形，面向风的方向，为茶厅提供有效的通风。不能开启的窗

d. 图书馆内部
e1./e2. 图书馆入口

d. Interior of the library
e1./e2. Entrances of the library

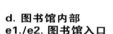

side of the site. Today, the entrance terrace, and the mighty screen including the "light and wind port" of the library, close off the square. The floor plan clearly reveals how the walls have been rotated out of the orthogonal geometrical pattern. These walls have been oriented into the wind (west to east), and modulate through three round windows the incident light admitted to the three-story reading room of the library. The feature of spatial depth zones and dual walls is clearly pronounced in this building. The wing walls set at angles are actually brisesoleils: not separate elements here, but integral constituents of the space as a whole. Kahn solved the problem of diffraction of the brilliant sunlight in a highly sophisticated manner in this building. The round openings, seemingly punched-out, have become gigantically dimensioned parts of the architecture. The circular segments in brick are held in place by prefabricated concrete joists. Kahn amalgamates traditional brick with modern concrete to achieve a harmonic entity. The round windows in the library were originally open, but have since been glazed. This measure was a necessity, owing to powerful monsoon winds and to the coming and going of doves although the interior has lost much of its archaic impact.

The dormitories were designed such that the side walls of the two wings form a funnel, facing the wind, which provides effective ventilation of the Tea Hall.

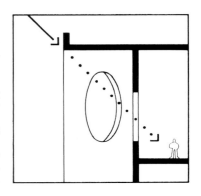

e2.

户下方可调节的通风百叶可以控制室温。底层容纳了公共活动室，每个公共活动室上方是五间学生寝室，沿着三角形平面的两边布置。房间前面的阳台打散了直射阳光。西立面上巨大的圆形洞口带来了穿堂风，也能通过这个洞口从公共区域望向阴凉的内庭院。厨房和服务用房相连，朝西，像一个背包。半圆形的楼梯位于三角形平面的焦点处，像灯塔一般控制着整组建筑。

宿舍和学校建筑群之间通过一条坡道和桥相连。路易斯·I. 康在他的设计图上写道："所以这个系统基本上是开放的。外部照射着阳光，内部则是你生活、工作和学习的地方。避免使用遮阳板这类构件，代之以深深的门廊，门廊中有凉爽的阴影。"（24）

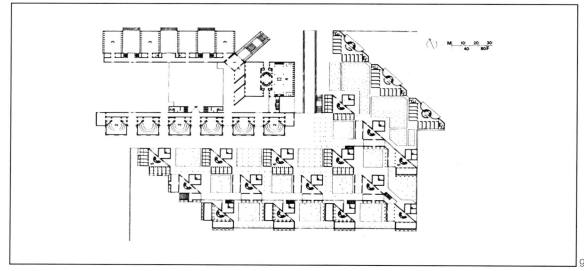

f. 窗户细部
g. 总平面图
h. 平面图
i. 宿舍

f. Window detail
g. Site plan
h. Plan
i. Dormitories

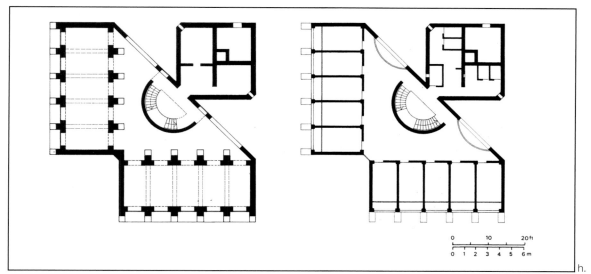

Adjustable ventilation panels below the non-opening windows permit regulation of room temperature. The ground floor contains the common rooms, above each of which are five students' rooms, arranged along two sides of the triangular floor plan. Balconies arranged in front of the rooms diffract the directly incident sunlight. The west facade contains huge round openings which provide cross-ventilation and allow a view from the common area into the shady inner courtyard. The kitchens and service rooms are connected, facing the west, like a backpack. The semicircular stairway, located at the focal point of the triangular floor plan, dominates the entire building complex as light tower.

A ramp and bridge connect the dormitories to the school complex. Louis I. Kahn writes on his design: "So the system is fundamentally that of open porches. The exterior is given to the sun and the interior is where you live and work and study. The avoidance of devices like brisesoleils brought about the deep porch which has in it the cool shadow." (24)

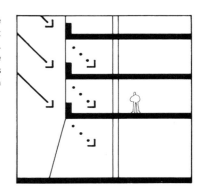

美国风交响乐团驳船
1965–1975；建成
匹兹堡，
宾夕法尼亚州

BARGE FOR THE
AMERICAN WIND
SYMPHONY
ORCHESTRA
1965 – 1975; BUILT
PITTSBURGH,
PENNSYLVANIA

路易斯·I.康为匹兹堡美国风交响乐团设计过两艘"音乐会驳船"。第一艘比较端庄的版本于1961年在伦敦投入使用，并在第一个音乐会季之后拆掉了。康设计的第二艘船的整体轮廓像一艘煤炭驳船，配备三台柴油发动机和海上作业所需的导航设备。自1975年完工后，这艘驳船已在美国广大的河流和运河系统上的农村地区停泊演出。甲板上有舞台、隔间和一间小演讲室。船体还提供了艺术展览空间。
设计为声罩的屋顶可以通过两个液压油缸在管弦乐队舞台上方移动。航海舷窗的形式为康设计这艘驳船的外观提供了灵感。略弯曲的钢板结构下容纳了演讲室和其他空间。康在早期设计中使用了圆形洞口的母题（参阅印度管理学院）。如康一直以来一样，此处采用的基本原则是：在细分的空间结构前，设置无玻璃的巨大构件来遮蔽阳光。在外置墙体与使用空间之间的空间来调节射入的光线。从驳船向外望去的景色被框在圆形的洞口中，令人印象深刻。

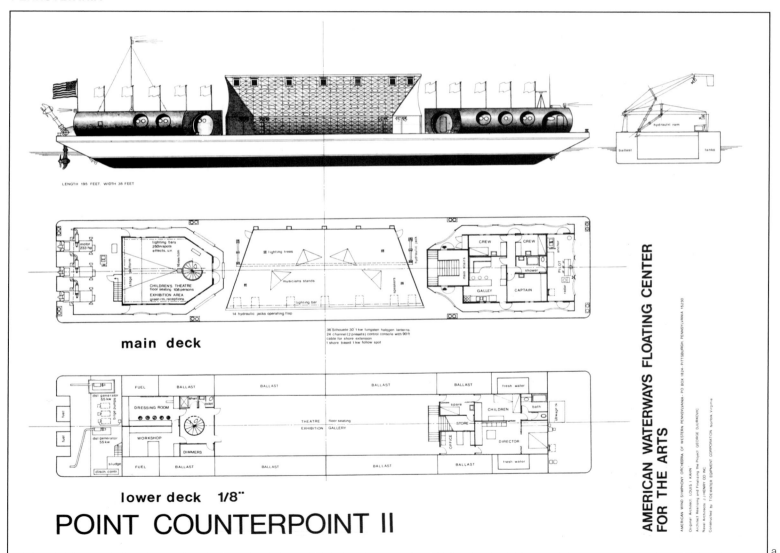

a.

Louis I. Kahn designed two "concert ships" for the American Wind Symphony Orchestra of Pittsburgh. The first, modest version was put into service in London in 1961 and disassembled after the first concert season. The second ship which Kahn designed resembles in overall silhouette a coal barge, and features three diesel motors and all the navigation equipment required for full maritime operation. Since its completion in 1975, this barge has made concert stops in rural areas on the extensive system of rivers and canals in the USA. The deck accommodates the stage, cabins, and a small lecture room. The hull offers space for art exhibitions.

A roof designed as acoustic shell can be moved by two hydraulic rams over the orchestra stage. The form of a nautical porthole served as inspiration for Kahn in his design of the exterior skin of the ship. A slightly curved sheet-steel construction accommodates the lecture room and other spaces. Kahn had in earlier designs used the motif of circular openings (cf. Indian Institute of Management). As always with Kahn, the basic principle involved here is: provision of formal macroconfiguration for a nonglazed wall situated for the purpose of sunlight protection in front of finely structured space. The space between the fore-placed wall and the usable room serves for modulation of incident sunlight. The view outward from the barge is impressively framed by the round cutouts.

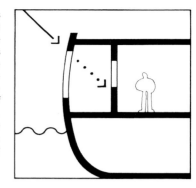

a. 立面图、平面图及剖面图
b. 走廊
c. 驳船

a. Elevation, plans, and section
b. Passageway
c. Barge

135

菲利普斯·埃克塞特学院图书馆
1965–1972；建成
埃克塞特，新罕布什尔州

PHILLIPS EXETER
ACADEMY, LIBRARY
1965 – 1972; BUILT
EXETER, NEW HAMPSHIRE

"图书馆里的眩光很不好；墙体空间很重要。你能够带着一本书在那里休息一会儿的小空间也非常重要……"（25）用康的话来说，这是图书馆中三个最重要的先决条件，他承认在埃克塞特学院图书馆中满足了后两个。然而，我不敢说他是否成功地解决了第一个问题。图书馆的四个立面结构完全相同；入口在北侧。这幢非常壮观的建筑由砖和混凝土建成，严格按照古典建筑的原则，分为基础、主体和檐部。外墙柱子的收分也给人留下了深刻印象。柱间两层高的窗户下部是木质的。成对的小窗户照亮了阅读壁龛；此处，读者可以关上滑动嵌板，阻隔掉公园景色的干扰，全心投入阅读当中。阅读桌上方是大面积的玻璃窗，没有窗挺，完全暴露在阳光下。

"原本，我在大窗户上设计了百叶。我仍然觉得在某些立面装上百叶会更好。北面不需要，但装在西面会非常有用。之所以没有装是因为它们可以随时轻松安装，而且当时也快没钱了。窗帘也同样是个不错的解决方案。"（26）

a.
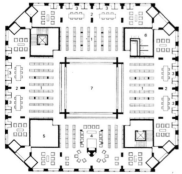

b.

"Glare is bad in the library; wall space is important. Little spaces where you can adjourn with a book are tremendously important...." (25) These are, in Kahn's words, the three most important prerequisites for a library of which he admittedly fulfilled the last two in Phillips Exeter Library. I hesitate to say, however, whether he achieved a successful solution for the first. All four facades are identically structured; the entrance is on the north side. This highly imposing building, executed in brick and concrete, strictly follows the classical architectural rules for base, main order, and attic. The upward tapering of the piers of the external wall is also impressive here. The two-story windows, installed between the pillars, feature lower sections made of wood. The pairs of small windows illuminate reading niches; here, the reader can shut the sliding panels to close off the distraction of the view into the park, in order to concentrate fully on reading. Above the reading tables are huge sections of glass, installed without muntins, which are fully exposed to the sun.

"Originally, I had shutters on the larger windows. I feel still that shutters would be good on certain elevations. They didn't have to be on the north, and possibly would have been most advantageous on the west. This was not done because they can be easily installed at any time, and money was running out. The blinds were an equally good solution." (26)

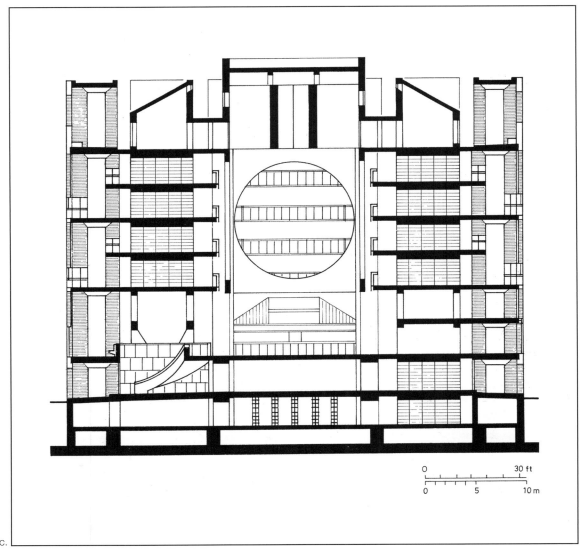

a. 平面图
b. 西南立面
c. 剖面图

a. Plan
b. Southwest facade
c. Section

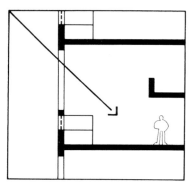

剖面图清晰地展示了由外向内的深度分区结构：在外部，两层通高的边缘区域是阅览空间，是砖砌的；然后是混凝土浇筑的部分；最后是混凝土核心部分，是一个中庭，墙上有四个巨大的圆形洞口。交叉梁对角横跨在正方形空间上方，用来将两侧射入的阳光向下反射，并在这个空间中形成向上升腾的特质。
"在中庭中，我选择了遮蔽光线的结构，这样光线就不会倾泻而下。"（27）
通过耶鲁大学美术馆的楼梯间，我们了解到康的"修光片"原理，即将光线向下反射，同时起到结构作用。

对角布置的交叉梁不仅强调了中庭的中心，而且在内部使得主体和檐部之间的界限清晰可见。
对混凝土板上切出的四个巨大圆洞的最好诠释，不是望向风景的窗，而是通往书的世界的窗。因此康的设计是在邀请读者去挑选自己的书籍，并带着它到一个阅读壁龛，或是图书馆的壁炉处去阅读。

d2.

d1.

The section clearly reveals the depth-zoned configuration from the exterior inward: on the outside, the double-story edge zone of the reading area in brick, then the area of the stacks in concrete, and last the concrete core configured as an atrium provided with four huge circular cutouts in the walls. The cross diagonally spanning the square inner space serves to conduct the laterally incident light downward and to furnish an upward definition of this space. "In the central room I chose the kind of structure which shields the light so it is not pouring down." (27)

From the stairwell tower of the Yale University Art Gallery we are familiar with Kahn's principle of "light blades" which deflect light downward and simultaneously perform structural functions. The diagonal position of the cross not only emphasizes the center of the atrium, but also renders definitely visible on the inside the border between main order and attic.

The four large circles cut out of the concrete slabs are best interpreted as windows not onto landscape, but into the world of books. Kahn's design therefore proves inviting to readers to select their book and take it into a reading niche or to the library fireplace to read.

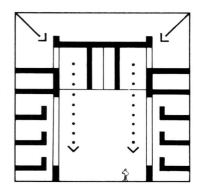

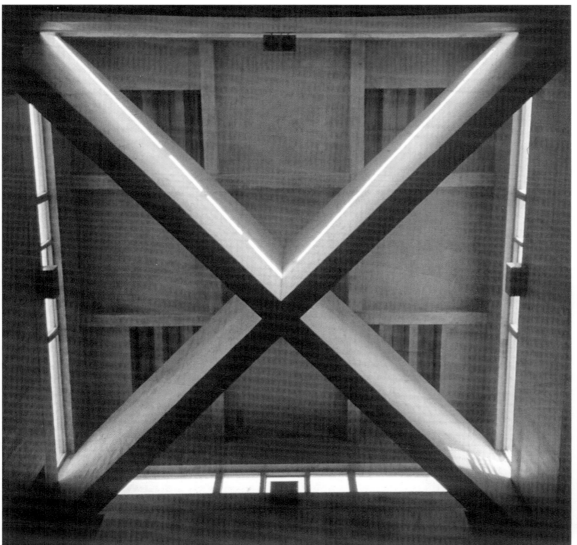

d1./d2. 阅读壁龛
e. 中央空间
f. 修光片

d1./d2. Reading niches
e. Central space
f. Light blades

奥利维蒂—安德伍德工厂
1966–1970；建成
哈里斯堡，
宾夕法尼亚州

OLIVETTI-UNDERWOOD
FACTORY
1966 – 1970; BUILT
HARRISBURG,
PENNSYLVANIA

"在奥利维蒂工厂，屋顶是有光的屋顶。看起来光线就像是穿透了屋顶。好像它并没有被覆盖。光线必须靠近工作台……一切都是开敞的。奥利维蒂工厂的窗户也在你的上方。"（28）

开放空间的设计是基于一个蘑菇式的结构。柱子依照正方形网格排列。相邻的八边形区域形成正方形的天窗，照亮了室内空间。着色玻璃纤维的金字塔形屋顶确保光线均匀入射。

立面由垂直的混凝土砌块墙体和连续垂直窗带交替排列组成。这些玻璃窗从地面一直延伸到混凝土蘑菇的下边缘，并采用由黑色细钢网薄片组成的构件遮挡直射阳光。用这种材料冲压而成并弯向水平方向的遮阳板相互紧密安装在一起，形成一个细丝网百叶遮阳板。然而，来自天窗和侧窗光线之间的相互作用在外立面上并不一致：外立面上的两种光线发生了直接冲突。

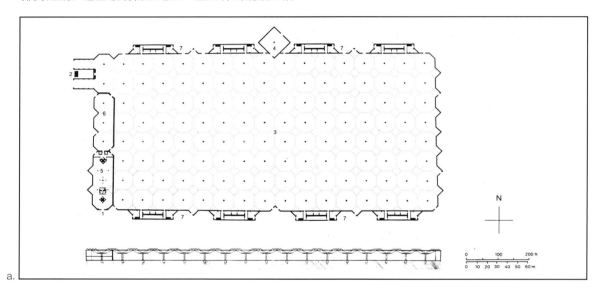

a.

b.

"In the Olivetti plant, the ceiling is a ceiling of light. Light looks right through it. It isn't as through it were covered. The light must be close to the work.... It is all open to view. Also the window in the Olivetti plant is above you." (28)

Design of the open-space room is based on a mushroom-construction principle. The columns stand in a square grid pattern. The contiguous octagon fields fit together to produce the square skylight which illuminates the interior. Tinted pyramidal roof elements of fiberglass ensure uniform incidence of light.

The facade is composed alternately of vertically configured concrete-block walls, and continuous vertical window strips. These glazed surfaces extend from the ground to the bottom edge of the concrete mushrooms and are shaded from direct sunlight by elements consisting of thin sheets of fine-meshed, black-coated steel. The sunshade panels stamped out of this material and bent to the horizontal are installed at close intervals to each other and go to make up a filigree brisesoleil. The interplay between light sources from the skylight and the windows is, however, not consistent at the periphery of the facade shell: there, two light sources directly conflict.

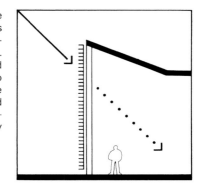

a. 平面图和剖面图
b. 东立面
c. 窗户细部

a. Plan and section
b. East facade
c. Window detail

KIMBELL ART MUSEUM
金贝尔美术馆
1966–1972；建成
沃斯堡，得克萨斯州

1966 – 1972; BUILT
FOR WORTH, TEXAS

"我的脑海中充满了罗马的伟大，拱顶在我脑海中刻下了深深的烙印，尽管我不能使用它，它却一直在那里。拱顶似乎是最好的。我意识到，光线必须从最高点射入，那里的光线是最好的。"（29）

路易斯·I. 康从一开始就确信这个美术馆应该是纵向布置的。到1969年，他已经为这个项目设计了四个版本。平面是矩形的，深深的切口作为入口区域。在西侧，美术馆与一个公园毗邻，边缘处有两个开敞的拱筒。对这个结构的第一印象是一个基于单一元素的重复形成的简单结构。北侧和南侧立面几乎没有透露出一点内部的壮丽。底层巨大的混凝土基座以及无窗的外壳给人留下了整体的印象。然而，实际上结构设计从外部清晰地展现出来：混凝土浇筑的结构柱支撑着上部的拱筒，非承重的填充墙则是由石灰华筑成的（与屋顶之间通过一条窗带分隔开）。拱筒是结构工程的杰作。每个拱筒都由四根柱子支撑，跨度约30米（104英尺），底部用作托梁。在拱筒顶部，有一条通长的槽形开口，使光线射入内部。康安装了横杆，以确保在最大压力和扭矩下结构的稳定性。边缘部分能够随着位置逐渐升高、

a. 光影变幻
b. 屋顶鸟瞰图
c. 底层平面图

a. Play of light
b. Aerial view of the roof
c. Ground floor

"My mind is full of Roman greatness and the vault so etched itself in my mind that, though I cannot employ it, it's there always ready. And the vault seems to be the best. And I realize that the light must come from a high point where the light is best in its zenith." (29)

Louis I. Kahn was convinced from the very beginning that this museum should be built with a longitudinal arrangement. By 1969 he had produced four project versions. The floor plan is rectangular, with a deep niche cut out for the entrance area. Toward the west, the museum borders onto a park with its two open cycloid vaults at the periphery. The first impression gained from this construction is that of a simple structure based on repetitive implementation of one element. The north and south facades provide little hint of the magnificence in the interior. The massive concrete footing of the ground floor as well as the windowless external shell create a monolithic impression. Nevertheless, the structural design is in fact clearly evident from the outside: the supporting pillars cast in concrete, with the cycloid vaults resting above, as well as the non-loadbearing infilling walls consisting of travertine (and separated from the roof by a strip of windows). The cycloid vaults represent a structural engineering masterpiece. Each of these vaults rests on four pillars, and spans a length of approx. 30 m (104 feet), with the base sections serving as joists. At the crown of the vault, a slot-shaped opening has been made the entire length of the structure, to

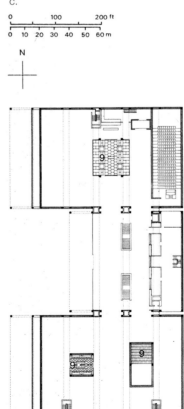

剪力逐渐增大承担更大的荷载；因此，拉杆不是必须的。奇怪的是，此处并没有将荷载标绘在平面详图上。拱筒是用光滑的模板现场浇筑的。混凝土内表面只是进行了清洁，保留了原本的粗糙。采石场的切割痕迹也保留在石灰华的表面，没有进行任何处理。

康在草图中进行了无数尝试，想要设计一种构件能够安装在拱筒顶部的采光槽下方，使光线能够在房间中均匀分布。他和一位助手经过长时间探索后终于找到了解决办法：通过50%打孔的弯曲铝制构件使光线重新分布，获得了想要的光线，其几何设计是根据太阳入射角度科学计算出来的。在建造过程中，建造了这个构件的全尺寸模型，并测试了其有效性：最终的构件理想地反射并分布了阳光。

"这个'自然采光构件'……是一个新的称呼方式；完全是一个新的词汇。它实际上是一个有效的调光器，能够在不同程度上控制光线的不利影响。当我看着它的时候，我真的觉得这是一件了不起的东西。"(30)

详图显示康如何精确地设计了"自然采光构件"的每个部分。拱筒顶部有机玻璃制采光口的设计尺寸，使

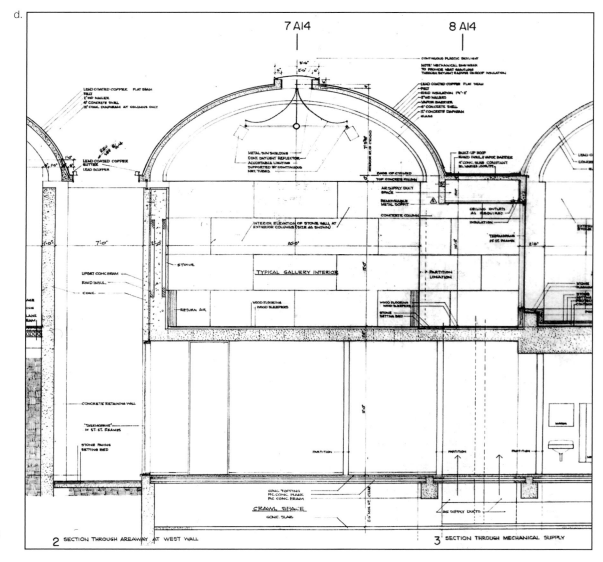

d. 剖面图
e. 自然采光构件细部
f. 地下室平面图

d. Section
e. Detail of natural light fixture
f. Basement

144

admit light into the interior. Kahn installed crossbars to ensure stability at this point of maximum pressure and torsion. The edge sections become capable of supporting more loads at progressively higher points, in accordance with the increase of shear forces; as a result, tension rods are not necessary. Strangely, the plot of forces has not (yet) been entered on the detail plan here. The cycloid vaults were cast as concrete onsite, with smooth formwork. The interior surface of the concrete is only washed, and otherwise left rough. The cutting marks from the quarry have also been left untreated on the travertine surfaces.

Kahn attempted in countless sketches to develop an element which could be installed under the light slot in the crown of the cycloid vault to uniformly distribute the light throughout the room. He and one of his assistants found the solution after a long search: the desired light distribution was successfully provided by 50% perforated, curved aluminum elements with geometrical design calculated scientifically according to the angle of incidence of the sun. During construction, a full-scale prototype of this fixture was installed and tested for effectiveness: the final element ideally reflects and distributes the sunlight.

"This 'natural lighting fixture' ... is rather a new way of calling something; it is rather a new word entirely. It is actually a modifier of the light, sufficiently so that the injurious effects of the light are controlled to whatever

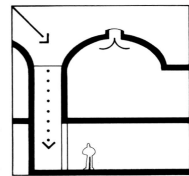

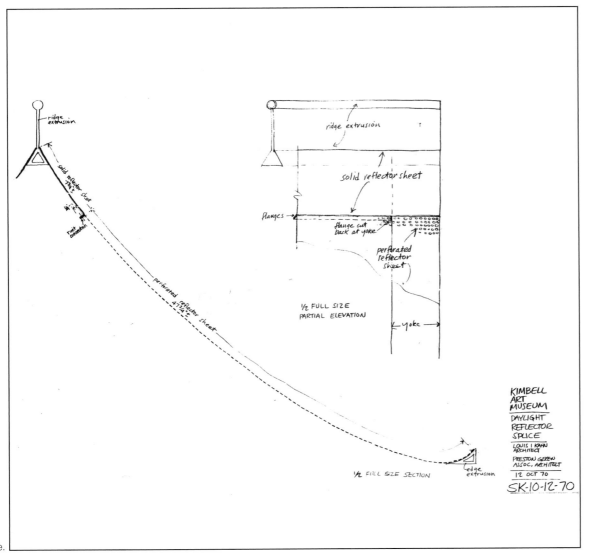

e.

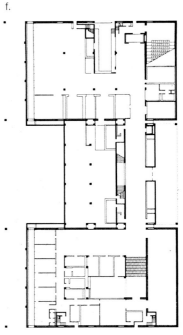

f.

一定量的光线落在两片修光片上,经反射后重新分布在空间中。因而房间不仅直接通过穿孔金属构件接收到光线,也接收到经过修光片反射到拱筒上的光线。"修光片"边缘安装有柔性聚光灯,补充人工照明。"建筑必须有对光的信仰。认为光是一切存在的给予者。每幢建筑、每个房间都必须有自然光,因为自然光给予一天的情绪变化。自然光将一年四季引入房间。甚至可以说,直到阳光落在建筑的墙面上,它才知道自己究竟多伟大。当光线进入房间,它就成为了你的光线,而不是别人的。它属于那个房间。金贝尔美术馆使用了所有自然光。"(31)

众所周知,康的自我评价:事实上,他在尝试从每个项目中获得新的见解。在他看来,金贝尔美术馆是自己最成功的建筑作品之一。

g. 内部
h. 自然采光构件细部

g. Interior
h. Detail of a natural lighting fixture

g.

degree of control is now possible. And when I look at it, I really feel it is a tremendous thing." (30)

The detail drawing shows how precisely Kahn designed each part of the "natural lighting fixture." The plexiglass-covered light opening at the apex of the cycloid vault is dimensioned in such a way that a particular quantity of light falls on the two blades for further distribution by reflection. The room therefore receives light not only directly through the perforated metal fixtures, but also indirectly as reflected from these blades and, in turn, from the cycloid vault. Flexible spotlights are installed at the edges of the "light blades" to provide additional artificial illumination. "An architecture must have the religion of light. A sense of light as the giver of all presences. Every building, every room must be in natural light because natural light gives the mood of the day. The season of the year is brought into a room. It can even be said that a sun never knew how great it was until it struck the side of a building. When a light enters a room, it is your light and nobody else's. It belongs to that room. The Kimbell Art Museum uses all natural light." (31)

Kahn's self-criticism was well-known: indeed, he attempted to gain new insights from each project. In his own view, the Kimbell Art Museum represents one of his most successful buildings.

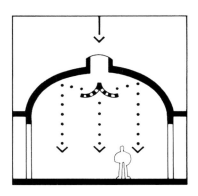

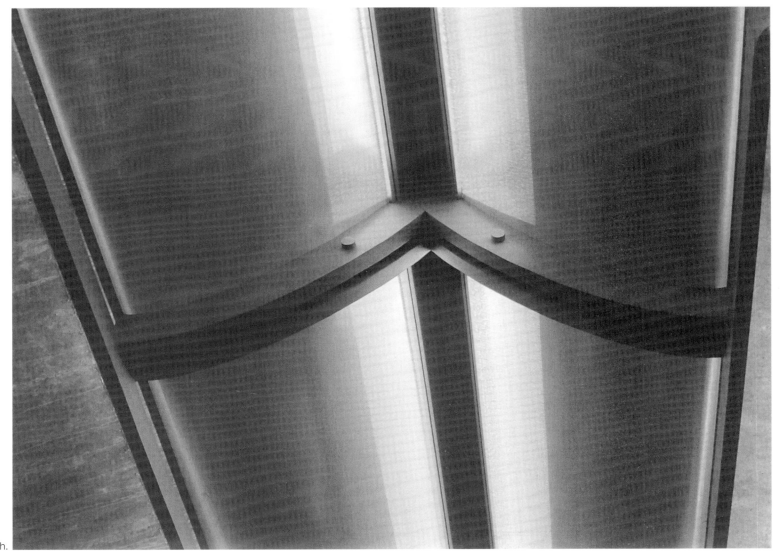

h.

贝塞尔犹太教会堂
1966–1972；建成
查巴克，纽约

TEMPLE BETH-EL SYNAGOGUE
1966 – 1972; BUILT
CHAPPAQUA, NEW YORK

康的设计是基于波兰和俄罗斯的木结构犹太教堂等历史范例。建筑平面呈八边形，四根混凝土柱子支撑着"采光塔"。康将讲堂围绕在犹太会堂四周，和在第一唯一神教堂中一样。然而在查巴克，是没有内走廊的：人们直接从中央大厅（至圣所）进入讲堂。其原因是：对于参与人数众多的活动，隔墙可以移到一边，使大厅容量从 250 人增加到 750 人。至圣所墙面完全是木质镶板，作为填充墙与混凝土框架结构融为一体。采光塔中有 24 个同样大的正方形窗户，全部没有窗挺。这座木塔的建筑表达及其窗户的规则性已经降到了最低。大量的窗户使至圣所能够感受到天气的变化。飘过的云朵、阳光、雨、薄暮，等等，直接决定着这个房间的特质，这是我在其他采用天窗照明的内部空间中从未体验过的。这种印象的形成可能是由于窗户与墙面齐平，于是它既不过滤也不调节光线。空调系统减弱了阳光在室内产生的热量。

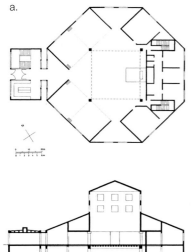

a.

b.

148

Kahn based his design on historical examples of synagogues executed in wood in Poland and Russia. The floor plan reveals an octagon shape, with four concrete columns supporting the "light tower." Kahn grouped the classrooms circumferentially around the synagogue chamber, as in the First Unitarian Church. In Chappaqua, however, there is no intervening corridor: one enters the classrooms directly from the central chamber (sanctuary). The reason: for well-attended events, the partition walls can be moved to the side to increase the capacity from 250 to 750. The sanctuary is paneled entirely in wood, which is integrated as infilling in the concrete skeleton construction. The light tower contains 24 identically large square windows, all installed without muntins. The architectural expression of this wooden tower, with the regularity of its windows, has been reduced to a minimum. The many windows allow the sanctuary to take part in changes of weather. Clouds moving past, sun, rain, twilight, and the like directly determine the character of this room to a degree which I had not experienced in other interior rooms illuminated by skylights. This impression may arise from the fact that the windows lie flush in the walls and therefore neither filter nor modulate the light. The air conditioning system compensates for the heat produced in the interior by the sun.

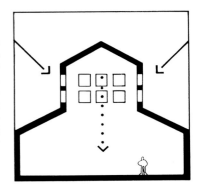

a. 平面图和剖面图
b. 南立面
c. 采光塔
d. 内部

a. Plan and section
b. South facade
c. Light tower
d. Interior

胡瓦犹太教堂
1967–1974；未建成
耶路撒冷，以色列

HURVA SYNAGOGUE
1967 – 1974; UNBUILT
JERUSALEM, ISRAEL

旧的胡瓦犹太教堂在 1948 年的战争中毁掉了。今天，巨大的石拱标志着它的原址。直到 1967 年，犹太社区才决定实施重建计划。然而，六日战争和赎罪日战争耗尽了必要的资金来源。

胡瓦犹太教堂是路易斯·I. 康设计的最伟大的作品之一，拥有强大的力量。石块垂直分层砌筑的桥塔，每座都有数吨重，环抱着内部的犹太教堂，如堡垒护卫一般，犹太教堂主体则是由轻质混凝土构件构成的。

"我感觉在犹太教中，烛光有着非常重要的作用。这些桥塔如蜡烛一般，有面向内室的壁龛。我觉得这是宗教本源的延伸，也是犹太教惯例的延伸。"（32）

在由 16 座桥塔组成的保护围墙之内，混凝土壳体组成的外壳形成向上张开的姿态。康对这个混凝土壳体背后的设计概念是这样描述的："这幢建筑的构成像是一棵树的大叶子，让光线滤过后进入室内。"（33）

值得注意的是，平屋顶略高于下部结构的边缘。如果从内部望向屋顶的下表面，屋顶板看起来就像是悬浮在那里，光线柔和地顺着壳体的曲面洒落下来。剖面图清晰地显示，康通过在犹太教堂核心处那些

b1.

b2.

a. 平面图
b1./b2. 总体模型
c. 模型

a. Plan
b1./b2. Site models
c. Models

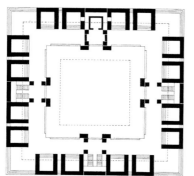
a.

The old Hurva Synagogue was destroyed during the 1948 war. Today, a large stone arch marks its original site. It was only in 1967 that the Jewish community decided to plan a reconstruction. The Six-Day War and the Yom Kippur War, however, drained the necessary financial resources.

Hurva Synagogue is one of the mightiest works in the sense of massive power conceived by Louis I. Kahn. Pylons consisting of vertically layered stone blocks, each weighing tons, constitute the protective bulwark around the inner synagogue, itself constructed of lightweight concrete components.

"I sensed that the light of a candle plays an important part in Judaism. The pylons belong to the candle service and have niches facing the chamber. I felt this was an extension of the source of religion as well as an extension of the practice of Judaism." (32)

Inside the protecting wall of sixteen pylons, the enclosure consists of a concrete shell which makes an upward-opening gesture. Kahn describes the concept behind the sweeping concrete shell as follows: "The construction of the building is like large leaves of a tree, allowing light to filter into the interior." (33) It is noteworthy that the flat roof is slightly lifted above the edge of the lower construction. If one looks up from the interior onto the bottom surface of the ceiling, the roof slab appears to float, and the light falls softly downward over the curved shell. The plan sections clearly reveal that

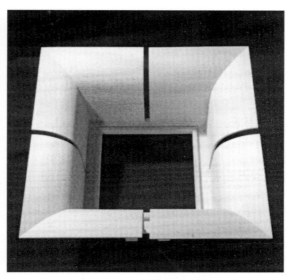
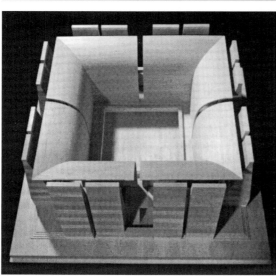

看似脆弱的混凝土结构外侧设计这些巨大的桥塔，实际上是实现了一个双重滤光器来调节光线。混凝土壳体像花朵一样向上张开，由 8 根相当细长的柱子支撑着，柱子布置在正方形的两条中轴线上。这种布置加强了康想要形成的印象：巨大桥塔内部的房间仿佛升腾起来，并开始盘旋。垂直缝隙将内部的混凝土壳分成四个部分：于是在一天中，有短暂的时间阳光会直射进入教堂，并且能最小限度地瞥见室外。犹太教堂在正方形的四角各有一个入口。一条环形回廊通向建筑内部。三部楼梯通往画廊。

康对胡瓦犹太教堂的光与空间的处理与他在达卡的作品之间显然存在某些相似之处。

对光线的特殊控制与空间质量、材料选择以及形式和结构特征使胡瓦犹太教堂成为路易斯·I. 康最美丽、最独特的设计之一。

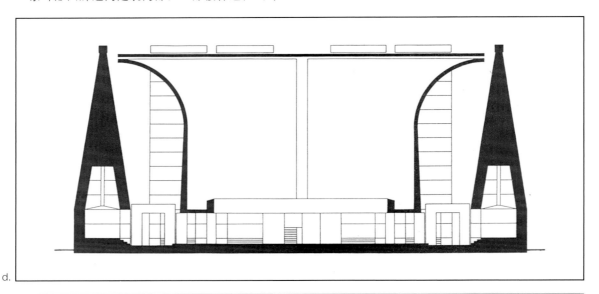
d.

e.

Kahn, by designing these massive protective pylons mounted in front of the seemingly fragile concrete construction at the heart of the synagogue, has in fact implemented a double filter for modulation of light. The concrete shell, opening upward like a flower, rests on eight relatively slender pillars arranged at the two center axes of the square. This configuration reinforces the impression which Kahn intended here: as though the room inside the mighty pylons has lifted and begins to hover. Vertical slots separate the inner concrete shell into four segments: as a result, direct sunlight falls for short periods of the day into the synagogue, and minimal outward glimpses are possible. The synagogue has four entrances one in each of the corners of the square plan. A circumferential ambulatory leads toward the inside of the building. Three stairways provide access to the gallery. Kahn's treatment of light and space for Hurva Synagogue demonstrates certain likeness to his work in Dhaka.

The special control of light in conjunction with the spatial qualities, the materials selected, and the formal and structural features make Hurva Synagogue one of the most beautiful and most distinctive designs conceived by Louis I. Kahn.

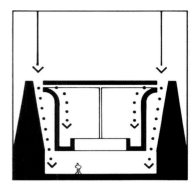

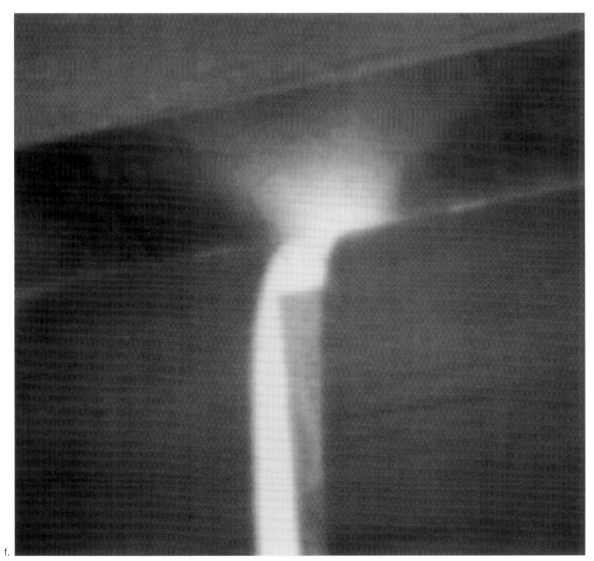

f.

d. 剖面图
e. 模型
f. 光带

d. Section
e. Model
f. Light pattern

沃尔夫森机械与运输工程中心
1968–1977；建成
特拉维夫市，以色列
继任建筑师：
摩西利—爱尔达建筑设计事务所

WOLFSON CENTER FOR
MECHANICAL AND
TRANSPORTATION
ENGINEERING
1968 – 1977; BUILT
TEL AVIV, ISRAEL
SUCCESSOR ARCHITECTS:
J. MOCHLY-I. ELDAR, LTD.

特拉维夫大学校园中的沃尔夫森机械与运输工程中心是康在以色列唯一实际建成的建筑。

康将长方形的机械大厅、教室、办公室以及自助餐厅成组布置，环绕在庭院周围。他将礼堂作为建筑群内部一个独立的建筑。东侧的两个附属建筑使建筑设施更加完备。康的三个大厅显然会令人想起他的金贝尔美术馆设计。康在这些大厅上方采用了类似的拱筒屋顶，并在顶部设计了采光槽。如我们在沃斯堡看到的一样，光线从屋顶上的窄槽射入，被"修光片"打散分布在整个室内空间中，尽管在沃尔夫森采用的技术方式更加简单一些。在特拉维夫，自然采光构件是由安装在带有加强肋的金属框中的铝条组成的。铝条安装精确，为了符合叶片的整体曲率。入射光被100%反射到拱筒屋顶上。人工光源安装在反射构件下侧顶点处。

康还在教室前设置了板墙，从而使耀眼的阳光间接进入室内。然而最终证明，从窗户射入的自然光是不够的，结果这些房间必须一直采用人工照明。

a.

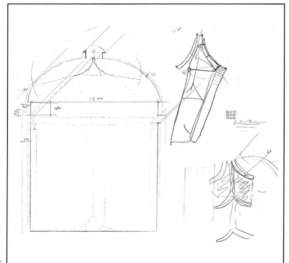

b.

c.

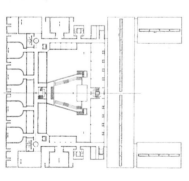

d.

The Center for Mechanical and Transportation Engineering on the university campus in Tel Aviv is the only building which Louis I. Kahn actually built in Israel.
Kahn grouped the extended oblong mechanical hall, the classrooms, the offices, and the cafeteria around a courtyard. He placed the auditorium as a lone building in the interior of the complex. Two annex buildings at the east of the site complete the facilities. Kahn's three halls here are strikingly reminiscent of his design for the Kimbell Art Museum. Kahn has likewise roofed these halls with cycloid vaults containing a light slot along their crowns. Light falling through the slot in the roof is distributed by "light blades" throughout the interior, as we have already seen in Fort Worth although the technical solution for Wolfson is simpler. In Tel Aviv, the natural lighting fixtures consist of aluminum strips mounted in a metal frame with reinforcing ribs. The aluminum strips have been precisely mounted to coincide with the required curvature of the overall blades. The incident light is reflected 100% onto the vault of the roof. Sources of artificial light are mounted at the underside apex of the reflector assembly.
Kahn also installed slab walls in front of the classrooms to admit the dazzling sunlight indirectly into the rooms. The windows and the admitted natural light proved to be insufficient, however, with the result that the rooms must always be artificially illuminated.

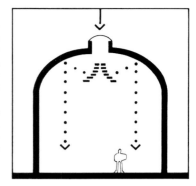

a. 平面图和剖面图
b. 设计草图（晚期版本）
c. 修光片细部
d. 屋顶鸟瞰图
e. 内部
f. 西立面

a. Plan and section
b. Sketch (late version)
c. Detail of light blades
d. Top view of the roof
e. Interior
f. Westfacade

耶鲁英国艺术中心
1969–1975；建成
纽黑文市，
康涅狄格州
继任建筑师：
米歇尔·戴维·迈耶斯和安东尼·佩莱基亚

从两方面来讲，梅隆中心（译注：此处作者用梅隆中心指代耶鲁英国艺术中心）都是路易斯·I. 康晚期作品中一个非同一般的项目。首先，他在这里采用了纯粹的混凝土框架结构，其次，他第一次成功地采用不锈钢板作为面层。可以肯定的是，在早期作品中，康曾尝试采用这种新的建造方式来探索自己的设计。他最初在美国劳工联合会医疗服务楼的尝试，被概念模糊所困扰，最终他在纽黑文的设计中厘清并解决了这些问题。在耶鲁中心，康采用了规则的柱网，并用中空的楼板取代了医疗服务楼中采用的空腹托梁加楼板的形式，并且楼板以完整厚度呈现在外立面上。在楼层之间，能看到清晰的由不锈钢板和挑出的滴水构件组成的三明治式结构。建筑平面分成了 60 个正方形区域，围绕着两个中庭布置。最初，正方形入口庭院设计成开敞的；然而后来由于气候原因，给它加上了屋顶。在第二个更大的中庭中，有一个独立的圆柱体作为垂直入口。这个楼梯与耶鲁大学美术馆楼梯之间的相似性是显而易见的，尽管在耶鲁中心，这个构件是安装在开放空间中的。光线穿过水平安装在混凝土圆柱体屋顶

a.

In two respects, the Mellon Center represents an exceptional project in Louis I. Kahn's late work. First, he executed the center as a pure concrete skeleton construction and, secondly, he for the first time successfully implemented stainless-steel plates as external cladding. Early in his work, to be sure, Kahn had attempted to employ this new mode of construction to develop designs of his own. His initial attempt, with the American Federation of Labor Medical Services Building, was plagued by conceptional ambiguities, problems which he definitively clarified and solved in New Haven. At the Yale Center, Kahn employs a regular column grid and replaces the Vierendeel joists plus floor slab as used for the Medical Services Building with floor slabs which contain interior cavities, and which appear at the facades on the exterior in their full thickness. Between the story levels, the sandwich constructions are visible, with stainless-steel slabs and protruding water-drip molds. The floor plan is broken down into sixty square fields, and is arranged around two atria. Originally, the square entrance court was planned to be open to the sky; later, however, on the basis of climate-control aspects, it was roofed over. In the second, larger atrium, a freestanding cylinder provides vertical access. The analogy to the staircase of the Yale University Art Gallery vis-à-vis is obvious, although this element is installed in open space at the Yale Center. Light comes through horizontally installed glass bricks in the roof slab of the concrete cylinder.

YALE CENTER FOR BRITISH ART
1969 – 1975; BUILT
NEW HAVEN, CONNECTICUT
SUCCESSOR ARCHITECTS: MARSHALL DAVID MEYERS AND ANTHONY PELLECCHIA

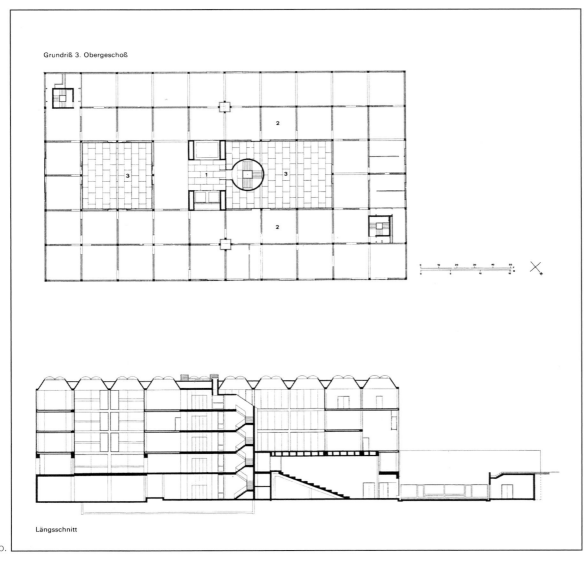
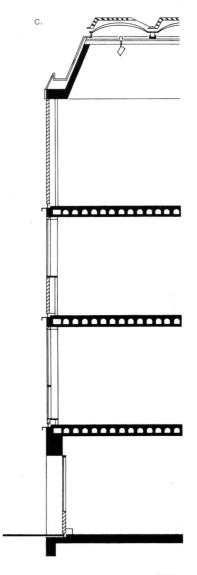

板下的玻璃砖照射进来。

在立面上，窗户与水平安置的钢板平齐。这种布置与混凝土柱和托梁共同形成了光滑的立面印象，仅露出向上逐渐缩小的混凝土构件和有角的滴水构件。康将水平滑动的木质百叶安装到墙壁结构中，以抵御眩光和直射阳光（参阅太阳屋和索克研究所的设计）。最初安装的是织物面板；然而在正式场合存在一些问题，因此换成了木质百叶。在提出为大面积的窗户遮挡直射阳光问题的 24 年后，路易斯·I. 康在此处实现了完美、协调的解决方法，采用滑动百叶，没有出挑的遮阳板，没有窗帘，没有特殊的染色玻璃，也没有突出的屋顶构件，等等。

覆盖整个屋顶的天窗，照亮了两个室内中庭和顶层的展厅。康在 58 个天窗圆顶上安装了复杂的遮阳构件，使得每年每天的每个时刻，都主要让北侧的光线进入室内。这些"遮阳帽"是康对一年中所有可能的入射阳光的角度进行细致研究的结果。为了获得关于光的质和量的精确数据，康搭建了这些遮阳构件 1:1 的木质模型，在旁边一个现有建筑屋顶上进行了试验。他原本设计了用柔性织物制成的可调节面板，

a. 东北立面
b. 平面图和剖面图
c. 剖面图
d. 墙体细部（晚期版本）
e. 内部
f. 滑动百叶

a. Northeast facade
b. Plan and section
c. Section
d. Detail of wall construction (late version)
e. Interior
f. Sliding panel

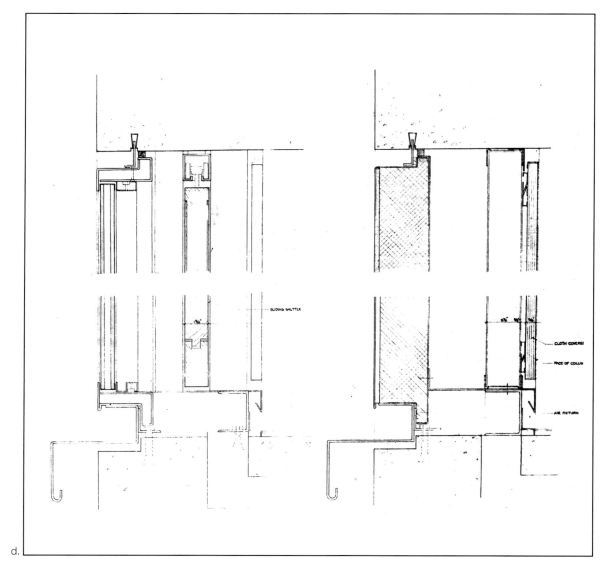

d.

In the facades, the windows are installed flush in the horizontally configured steel slabs. This arrangement together with the concrete pillars and the joists creates the impression of a smooth facade surface, structured only by the upward-tapering concrete members and the angular protrusion of the water-drip molds. Kahn integrates horizontally sliding wooden blinds into the wall construction to counter glare and directly incident sunlight (cf. design for the Solar House and the Salk Institute). Originally, fabric panels were installed: their use was problematic on formal grounds, however, and the wooden blinds were installed instead. Twenty-four years after he originally addressed the problem of protecting large windows areas from direct sunlight, Louis I. Kahn implements here the perfect, coherent solution, in the form of the sliding blinds without protruding briseso-leils, without curtains, without special tinted glass, without overhanging roof elements, etc.

Skylights, covering the entire roof, illuminate the two interior atria and the exhibition rooms on the top story. Kahn installed sophisticated sunshades on the 58 skylight domes which, at all times of the day and year, primarily admit north light into the interior. These "sunshade caps" are the result of painstaking studies carried out by Kahn on all possible angles of sunlight incidence during the year. In order to obtain exact data on light quantity and quality, Kahn conducted trials with 1:1 wooden mockups of these sunshades, mounted on the roof of an existing

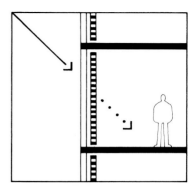

e.

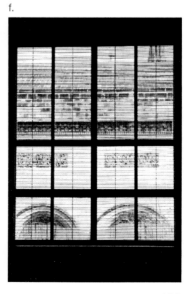

f.

以便更好地调节光线（参阅细部剖面），但是康最终放弃了这个方案，因为在天窗圆顶下，面板会产生空间问题。224个遮阳构件是由金属百叶组成的，由有机玻璃穹顶上的四根支柱支撑。这些构件在南侧形成相当大的遮阳面积，下面的遮阳百叶以能够最有效遮挡直射阳光的角度安装。为了确保令人满意的采光，有机玻璃穹顶由四个一组的"V"形混凝土托梁支撑（在内部的空腹中容纳着通风管道）。

天窗是由以下构件构成：两层波纹有机玻璃——一层带有反射涂层的光扩散层，一层透明有机玻璃用来防尘，内部是有机玻璃穹顶的紫外线吸收壳以及外部的天窗外壳。出于生产方面的考虑，要求将每个天窗单元分为四个穹顶，其中还包括用于人工照明的聚光灯。因此自然光和人工光融合成为一个兼具功能与形式的整体。

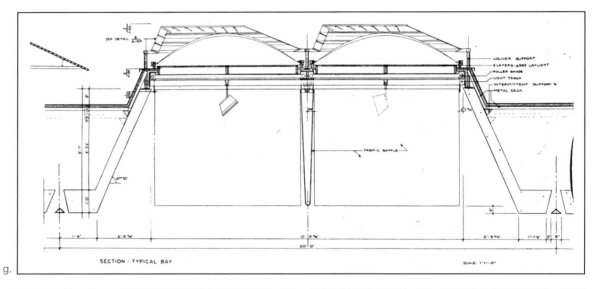
g.

h.

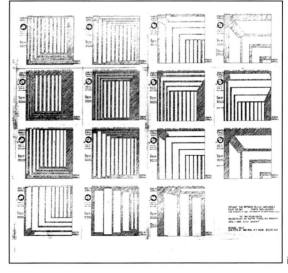
i.

g. 天窗剖面图（晚期版本）
h. 天窗
i. 阴影示意图
j. 屋顶

g. Section of skylight (late version)
h. Skylight
i. Shadow diagram
j. View of the roof

160

building next door. He originally planned for adjustable panels made of flexible textile to better control the light (see detail section), but Kahn discarded this solution since the panels would have created space problems under the skylight domes. The 224 sunshade elements consist of a metal louvre construction resting on four support posts above the plexiglass domes. These sunshades present a relatively large shading surface toward the south, and the brisesoleil louvres below have been installed at the most effective angles to the incident sunlight. To ensure satisfactory admission of the light, the plexiglass domes rest in groups of four on V-shaped concrete joists (which contain ventilation ducts in their hollow interiors).

The skylights consist of the following components: two layers of corrugated plexiglass, one light diffusor with reflector coating, one clear plexiglass layer as dust protection, the inner ultraviolet-radiation-absorbing shell of the plexiglass dome, and the exterior shell of the skylight. Production considerations dictated that four domes be grouped to each skylight unit, which also contains spotlights for artificial illumination. Natural and artificial light therefore blend to a functional and formal entity.

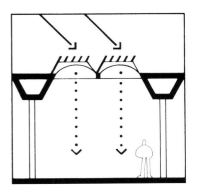

计划生育与妇幼保健中心
1970–1975；
部分建成
加德满都，尼泊尔

FAMILY PLANNING CENTER AND MATERNAL HEALTH CENTER
1970 – 1975;
PARTIALLY BUILT
KATMANDU, NEPAL

路易斯·I. 康在加德满都老王宫附近的一块三角地上规划设计了所需的公共设施。然而他的总体规划并未被采用。原本设计成"H"形的计划生育与妇幼保健中心建筑群，只有南翼最终建成。在加德满都，它被称为"没有屋顶的房子"：最初，康打算将露台作为屋顶花园。在康晚期的所有作品中，这个项目应用了最多的古典元素。三层通高的柱子，与嵌入其中的窗户单元，赋予建筑一种纪念性的气质，这种手法显然借鉴于鲍扎体系。绝对对称的楼层平面以及立面分为基础、主体和檐部三个部分，令人想起康在 1924 年学生时期的作品。壁柱在平面中清晰可见，其"U"形的平面用于容纳储物柜和垂直的水电管线。柚木制成的窗户凹进去，与柱子内侧对齐，这样可以遮阳挡雨。此处墙体的双重结构显而易见。窗龛底部的斜角以及开敞顶层的女儿墙，都使得柱子看起来更细更长。垂直的壁柱被滴水槽打断了。

a.

Louis I. Kahn planned and designed the layout for intended public facilities on a triangular plot near the old King's Palace in Kathmandu. His master plan, however, was not accepted. Of the original H-form design for the Family Planning Center complex, only the south wing was executed. It is called "the house without a roof" in Kathmandu: originally, Kahn had planned to execute the terraces as roof gardens. Of all Kahn's late works, this project is most heavily indebted to classicistic sources. The columns reaching over three stories, together with the inset window units, endow the building with a monumental character apparently borrowed from the terminology of the Ecole des Beaux-Arts. The absolute symmetry of the floor plan, as well as the facade breakdown into base, main order, and attic, call Kahn's student work of 1924 to mind. The pilasters clearly recognizable in the floor plan with their U-forms serve to accommodate cabinets and vertical water and power supply lines. The window elements, executed in teak, are recessed along the inner column line, where they are protected from rain and sun. The wall as double structure is clearly in evidence here. The oblique angle of the base section in the window niches, as well as the parapet of the open top story, both create the effect that the columns appear longer and more slender. The verticality of the pilasters is broken up by the brick water-drip molds.

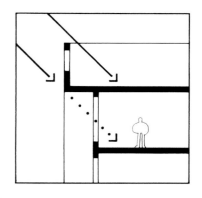

a. 南立面
b. 窗口
c. 平面图和剖面图

a. South facade
b. Window niche
c. Plan and section

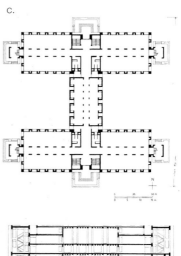

联合神学研究院公共图书馆
1971–1977；建成
伯克利，加利福尼亚州
继任建筑师：
埃希里克·霍姆希·多齐和戴维（EHDD 建筑事务所）以及彼得斯·克莱伯格和考尔菲尔德

COMMON LIBRARY,
GRADUATE
THEOLOGICAL UNION
(GTU)
1971 – 1977; BUILT
BERKELEY, CALIFORNIA
SUCCESSOR ARCHITECTS:
ESHERICK HOMSEY DOGE
AND DAVIS, AND PETERS
CLAYBERG & CAULFIELD

业主明确提出想要设计一个单一的大型图书馆，来满足天主教、新教及犹太教的需求。其目的是将不同信仰的学生团结在同一片屋檐下，为他们提供一个公共聚会场所。这个项目令人想起康设计的华盛顿大学图书馆：在伯克利，他同样采用阶梯金字塔的形式作为主题，但用一个"绿色过滤器"代替了华盛顿大学图书馆中的遮阳构件。这个过滤器由种植在窗前的树木组成：这是康的作品中第一次使用这样的理念。他原本想种橘子树，但出于气候考虑最终种植了 92 棵岛海桐和 26 棵海桐。在伯克利，阳光通过一个"采光井"进入室内。交叉排列的"修光片"的配置和构造是对康在埃克塞特学院图书馆采用的采光系统的进一步发展。然而由继任建筑师在伯克利实际完成的这幢建筑与康最初的设计完全不同。进一步说，如果按照最初的设计建造这幢公共图书馆，对于大进深带来的采光问题将会得到更令人满意的解决方案，针对这个问题，康在 16 年前的华盛顿大学图书馆设计中也曾因类似的形式而面临同样的困难。

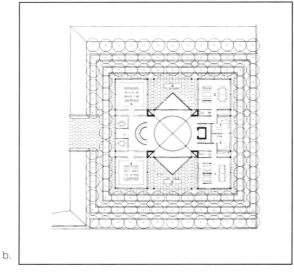
b.

c.

a. 剖面图
b. 平面图
c. 模型顶视图
d. 南立面
e. 东立面

a. Section
b. Plan
c. Top view of model
d. South facade
e. East facade

a.

d.

The clients specified that one large, single library be designed to serve Catholic, Protestant, as well as Jewish needs. The intention was to unite students of these religions under one roof and to offer them a common meeting place. This project calls to mind Kahn's design for the Washington University Library: in Berkeley, he likewise employs the stepped pyramid as dominant theme, but with a "green filter" instead of the sunshade elements planned for the Washington University Library. This filter consisted of trees planted in front of the windows: a concept employed here for the first time in Kahn's work. He originally planned to use orange trees, but climate considerations led to actual planting of 92 *Pittosporum undulatum* and 26 *Pittosporum tobira* trees. At Berkeley, sunlight is admitted to the interior through a "light shaft." The configuration and construction of the "light blades" arranged in a cross is a further development of Kahn's system employed for admitting light into Phillips Exeter Library. The building actually executed in Berkeley by the successor architects deviates extensively from Kahn's original design, however. With the Common Library as originally planned, furthermore, a more satisfactory solution would have been achieved for the light problems resulting from the great depth of this building, difficulties with which Kahn had already struggled in similar form sixteen years earlier, in his design for the Washington University Library.

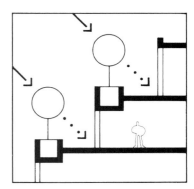

e.

孟加拉首都政府建筑群：医院
1962-1969；建成
达卡，孟加拉国

SHER-E-BANGLA NAGAR:
NATIONAL CAPITAL OF
BANGLADESH HOSPITAL
1962 – 1969; BUILT
DHAKA, BANGLADESH

a. 平面图
b. 剖面图
c. 西立面
d. 门廊

a. Plan
b. Sections
c. West facade
d. Portico

在最初设计的整个大型医院建筑群中，只有部分实际建成：砖砌入口部分、入口后面开敞的等候大厅以及检查和治疗楼的一部分。横跨整个医院长度的门廊由两个有深度的空间区域组成，它们反射阳光并提供阴凉。第一个深度空间区域设置了巨大的圆形洞口，垂直于外立面。天花板涂成白色，有助于漫射光线。原本立面上最上面的圆洞没有安装玻璃。砖砌圆洞中的混凝土托梁有助于突出门廊的端头部分和建筑的中间部分。康将建筑分成了偶数个部分（8个），以免入口设置在对称轴上（偏离中心）。

第二个深度空间区域是完全开敞的。密肋楼板跨过整个开敞的区域，从这里可以沿附属楼梯向左或向右走。一层楼的大会议室中，横向布置的混凝土遮阳板将阳光反射到天花板上，然后以漫射和间接的形式进入大厅内部。这种调光构件是首次出现在路易斯·I. 康的作品中，他再次向我们展示了他是如何不断地与光和光的力量进行概念上的斗争。

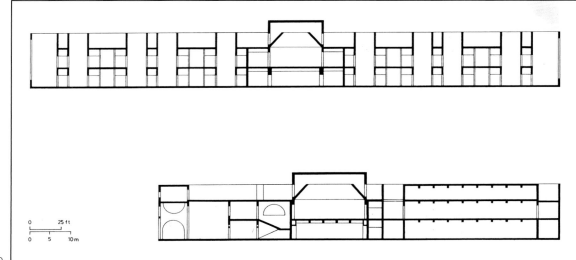
b.

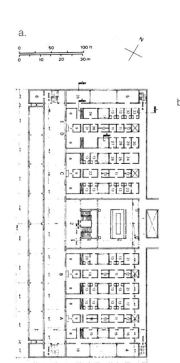

c.

166

Of the entire large hospital complex originally planned, only part was built: the brick entrance section, the open waiting hall behind the entrance, and the examination and treatment wing. The portico, which stretches over the entire length of the hospital, consists of two spatial depth zones which diffract the sunlight and provide cooling shade. Giant circular segments have been installed in the first depth zone, perpendicular to the outer facade. The ceiling has been plastered in white and therefore helps to diffuse the light. Originally, the uppermost circular segments in the facade were not glazed. The concrete joists in the brick circles help to highlight the end of the portico as well as the middle section of the building. Kahn broke down the building into an even number of subdivisions (eight), in order to avoid placing an entrance on the axis of symmetry (off-set center). The second spatial depth zone is open throughout. A ribbed floor slab spans the long open area, from which secondary flights of stairs lead off to the left and right. In the large meeting room on the first upper story, laterally arranged concrete louvers reflect the light onto the ceiling, from where it falls in diffused and indirect form into the hall interior. This type of light modulation appears for the first time in the work of Louis I. Kahn, who shows us here again how he constantly struggles with the concept of light and its power.

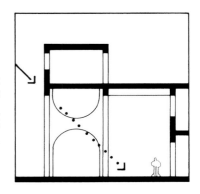

d.

孟加拉首都政府建筑群：
议会堂
1962–1983；建成
达卡，孟加拉国
继任建筑师：戴维·威兹德姆及合伙人事务所

1962年，路易斯·I.康受委托设计伊斯兰堡（当时的西巴基斯坦，现在的巴基斯坦）的行政首都以及达卡（当时的东巴基斯坦，自1973年起为孟加拉国）的立法首都。伊斯兰堡的规划工作1965年结束，但达卡项目的建设在康死后才完成。

议会堂是康最美、最成熟的作品。他在达卡采用了各种各样的光线控制和调节技术，我将研究范围限定在议会大厦和医院。餐厅和政府部长、秘书、助手和宾客的招待所采用的是与其他项目相似的技术：主要是印度管理学院。

"我得到的任务包括大量建筑项目：议会堂、最高法院、招待所、学校、体育场、外交飞地（责编注：外国使领馆区）、生活区、市场：所有这些布置在一块面积1000英亩遭受洪水侵袭的平坦土地上……我一直在考虑这些建筑应当如何组合以及是什么使他们在这片土地上占据一席之地。第三天晚上，我从床上下来，脑子里想着该规划的主要理念。我画下第一张草图，是规划了清真寺和湖泊的议会堂。我在湖周围加上招待所。仅仅是因为意识到'集会'具有至高无上的特点。人们聚集在一起感受社区精神，

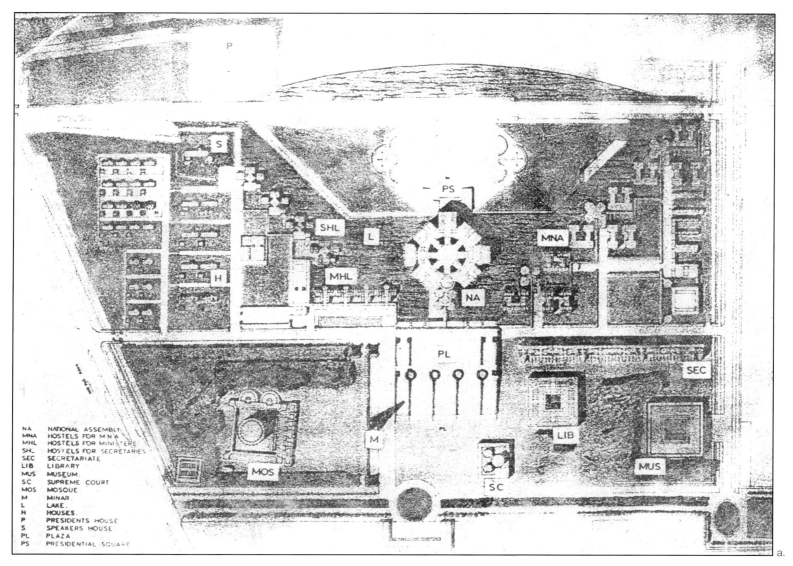

NA NATIONAL ASSEMBLY
MNA HOSTELS FOR MNA
MHL HOSTELS FOR MINISTERS
SHL HOSTELS FOR SECRETARIES
SEC SECRETARIATE
LIB LIBRARY
MUS MUSEUM
SC SUPREME COURT
MOS MOSQUE
M   MINAR
L   LAKE
H   HOUSES
P   PRESIDENTS HOUSE
S   SPEAKERS HOUSE
PL PLAZA
PS PRESIDENTIAL SQUARE

a.

In 1962, Louis I. Kahn was commissioned to design the executive capital in Islamabad (then West Pakistan, now Pakistan) and the legislative capital in Dhaka (then East Pakistan, since 1973 Bangladesh). The planning work was stopped in 1965 for Islamabad, but construction of the project in Dhaka was completed after Kahn's death.

The Assembly Hall is very well Kahn's most beautiful and most mature work. From the great variety of light-control and modulation techniques he employed in Dhaka, I have limited my consideration to the parliament building and to the hospital. The solutions employed for dining rooms, and hostels for government ministers, secretaries, aides, and guests are based on techniques familiar from other projects: primarily, the Indian Institute of Management.

"I was given an extensive program of buildings: the Assembly, the Supreme Court, hostels, schools, a stadium, the diplomatic enclave, the living sector, market: all to be placed on a thousand acres of flat land subject to flood.... I kept thinking of how these buildings may be grouped and what would cause them to take their places on the land. On the night of the third day, I fell out of bed with a thought which is still the prevailing idea of the plan. I made my first sketch on paper, of the Assembly, with the mosque and the lake. I added the hostels framing this lake. This came simply from the realization that "assembly" is of a transcendent nature. Men come

SHER-E-BANGLA NAGAR: NATIONAL CAPITAL OF BANGLADESH ASSEMBLY HALL 1962 – 1983; BUILT DAHKA, BANGLADESH SUCCESSOR ARCHITECTS: DAVID WISDOM AND ASSOCIATES

b.

a. 总平面图（晚期版本）
b. 南立面

a. **Site plan (late version)**
b. **South facade**

我觉得这必须是可表达的。通过观察巴基斯坦人宗教生活的方式，我认为将清真寺编织入议会堂的空间网络中能够表达这种感受。假定这个有些武断的观点是正确的。我怎么能知道这会适合他们的生活方式呢？但这个假设成为了一个支点。"（34）

第一张草图上包含了这个建筑群中所有的基本要素。议会堂和清真寺设在中心，政府部长和秘书的招待所布置在左右两翼，与主楼之间用人工湖隔开。此处展示的总平面图仅是晚期版本，因为很不幸最终版本不能发表。路易斯·I. 康很少使用"议会堂"这个表达方式：而是用"人类制度的堡垒"来代替（35）这幢建筑的确像一座巨大的堡垒。这幢建筑被水面包围，仅通过坡道和桥梁与大陆相连，使它更像一座防御工事了。

平面清晰展示了议会堂周围各种各样建筑体的形式。总统入口位于北角，有对角布置的楼梯。政府部长的房间朝西，清真寺朝南，向着麦加的方向。康将大餐厅和会议室布置在东边。四个办公体块插入这些主要的元素之间。一个环形的回廊从底层地面延伸到屋顶，环绕着中心对称的会议室。

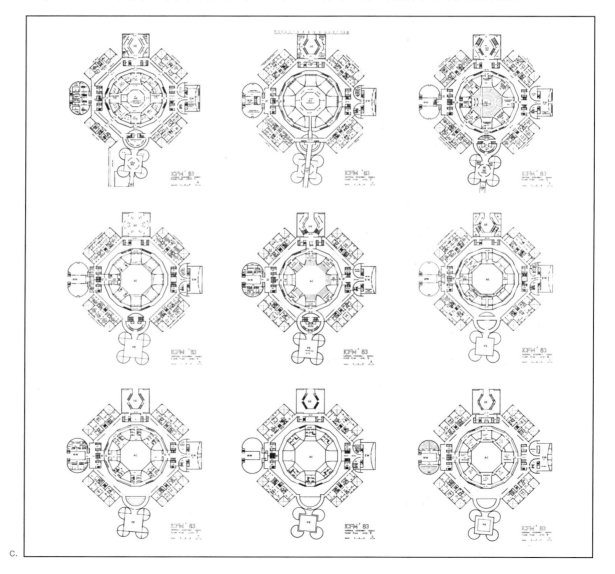

c. 平面图
d. 西立面
e. 回廊

c. Plans
d. West facade
e. Ambulatory

to assembly to touch the spirit of community, and I felt that this must be expressible. Observing the way of religion in the life of Pakistani, I thought that a mosque woven into the space fabric of the Assembly would reflect this feeling. It was presumptuous to assume this right. How did I know that I would fit their way of life? But this assumption took possession as an anchor." (34) This first sketch contained all the essential elements of the complex. The parliament and mosque are set in the center, with hostels for government ministers and secretaries arranged in wings to the left and right, and separated by an artificial lake from the main building. The site plan shown here is merely a late version, since the final plan cannot unfortunately be published. Louis I. Kahn rarely used the expression "the Assembly Hall": instead, he spoke of "the Citadel of the Institutions of Man." (35) The building indeed resembles a huge fortress. The fact that the building is surrounded by water and connected only by ramps and bridges to the mainland reinforces the impression of a fortification.

The plan clearly reveals the forms of the various building components around the Assembly Hall. The presidential entrance, with the diagonally arranged flights of stairs, is in the north corner. Prestigious rooms for government ministers face the west, and the mosque is toward the south, oriented to Mecca. Kahn located the great dining and meeting room to the east. Four office wings are inserted among these conspicuous elements.

d.

e.

一进入这个过渡区域，你就会为光影的壮丽所折服。阳光，透过屋顶的玻璃砖照射进来，像水流一样从墙面上淌下，进入内部空间。公共空间的特质和氛围随着一天中太阳的运行而变化。

密肋屋顶上嵌入的带状玻璃砖相互呈直角布置：这使得墙和地面上闪烁的光影变幻更加有趣。由于办公部分的骨架没有延伸到屋顶，光源和可见光之间的位移清晰可见：密肋楼板仿佛飘浮在空中。像是从回廊的侧墙上冲出的巨大半圆形、圆形和三角形洞口为走廊和通往议会堂和办公室的楼梯提供采光。由于这些洞口在阴影或逆光中，它们具有了更强的形式意义。

楼梯位于一个空间壳体中，插入底层环形通道和议会堂之间。这些楼梯在巨大的圆形洞口背后斜向展开，为上层的议会堂、小办公室、电视广播工作室及顶层的技术用房提供采光。内部通道通过栈桥与东西两侧的外部走廊相连。

八个楔形采光井穿过建筑的核心。它们的顶点界定了八边形的议会堂。这些内庭院有三面装了玻璃，从而为毗邻的房间提供采光（和通风）。它们的内

f.

g.

h.

i.

A ring-shaped ambulatory, extending from the ground floor to the ceiling, circles the centrally symmetric Assembly room.

Upon entering this intermediate zone, one is overcome by the magnificence in the play of light. Sunlight, entering through glass blocks in the roof, flows like a stream of water over the walls and into the interior. The character and the atmosphere of the common space change with the daily progression of the sun.

The strip-configured glass blocks in the ribbed roof are mutually arranged at right angles: which makes even more interesting the play of light and shadows on the walls and floor. Since the inner shell of the office complex does not extend to the ceiling, the displacement between the light source and the visible sunbeams is clearly perceptible: the ribbed ceiling appears to float. The huge semicircles, circles, and triangles seemingly punched out of the walls flanking the ambulatory provide illumination for the corridors and staircases leading to the Assembly Hall and offices. Since these openings lie in shadows or in counterlight, they are endowed with even greater formal impact.

The stairways are located in a spatial shell inserted between the access ring on the ground floor and the Assembly Hall. These stairways pass as diagonals behind giant round openings which provide light to the upper levels of the Assembly Hall, small offices, TV and radio studios, and technical rooms on the top story. The

f. 玻璃砖
g. 回廊
h. 采光井
i. 屋顶
j. 密肋楼板

f. **Glass blocks**
g. **Ambulatory**
h. **Light shaft**
i. **View of the roof**
j. **Ribbed floor slab**

部设计纯粹是功能性的。然而，建筑顶部八个巨大的垂直圆洞则使得顶层显得相当壮观。采光井顶部是开敞的。庭院内部的铝制窗框抵御了风雨的破坏。这种解决方式，即光的垂直入射，在费城艺术学院项目中也能够看到（1960—1966）。

两个办公部分是彼此镜像的，均围绕着正方形中庭"L"形排列。楼梯和前方的门厅是通过一整个楼通高的凹进"槽形窗"来采光的。采光井所在的每个转角处的墙面上都有一个矩形、直角三角形和一个圆形洞口。这些洞口带来了向外眺望的视野，并过滤了直射阳光。然而，主要光源是中庭屋顶上的玻璃砖。原本，康想要让这个空间向天空开敞。然而，由于强季风降雨及其对木质窗户构件的影响，1974年后决定用透明材料将采光井封闭起来。结果，这里的采光空间大幅失去了它原来所具有的建筑力量。

共有128块玻璃砖安装在正方形混凝土肋之间，发散了阳光，使办公空间中的光线更均匀。

墙上的水平石条标记着混凝土浇筑模板的高度，高度为5英尺（此处一层为10英尺，即大约3米）。混凝土浇筑带之间的接缝中嵌有白色大理石。在外

k1.

k1./k2. 回廊中的光影变换

k1./k2. Play of light in ambulatory

inner access way is connected by bridges to the outside corridor on the east and west.

Eight wedge-shaped light shafts penetrate to the core of the central building. Their apices define the octagon of the Assembly Hall. These inner courts are glazed on three sides and hence permit illumination (and ventilation) of the adjoining rooms. Their design at the interior is purely functional. The eight huge vertical circles at the top of the building, however, distinctively characterize the imposing attic level. The light shafts are open to the sky. The use of aluminum window frames in the interior of the courts eliminates damage by the elements. This solution, the vertical admission of light, can also be found in the Philadelphia College of Art project (1960 - 1966).

The two office sections are designed as mirror images of each other, and are each set in an L-arrangement around a square atrium. The stairway complex with the foyer in front is illuminated by a recessed "slot window" extending over all stories. Each of the corners in which the light shafts are located features a rectangle, a vertical triangle, and a circle on the side. These openings allow a view toward the outside, and filter the directly incident light. The main source of light, however, is the roof of glass blocks over the atrium.

Originally, Kahn planned to leave this space open toward the sky. Owing to the severe monsoon rains and their effects on wooden window components, however, the decision was reached after 1974 to close off the

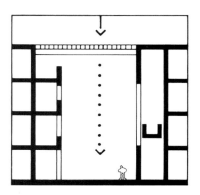

k2.

立面上，楼板的高度是通过滴水槽表现出来的。

在议会堂绝对中心对称的结构中，清真寺的几何形状略偏离了南北轴，为了准确地面向麦加。立方体内部由四个"采光圆柱体"围护着。

祈祷室的特点是绝对内向的，没有窥孔，除了从顶部洒下的光线外什么都没有。从外面看，带有入口的塔楼令人想起城堡。然而，从内部来看，祈祷室的内部设计非常复杂也很非凡。半圆形和圆形的布置与办公室不同，此处的采光井是向着天空敞开的。从而在清真寺内可以感受到风和天气的变化，随着一天时间的变化，阳光进入祈祷室中。飘过的云朵强烈影响着内部空间的特质。按照清真寺的传统，人工照明灯具挂得很低，在夜间营造一种更加柔和、更加亲密的氛围。祈祷室下面的门厅也通过四个"采光圆柱体"来采光。

首都政府建筑群设计的绝对核心是议会堂，这是国家立法机构的议会堂，四周环绕着"服务空间"：办公室、部长办公室、接待室以及电视广播工作室。平面是基于一个八边形，上面覆盖着伞状拱形屋顶。

1964 年，康写道：

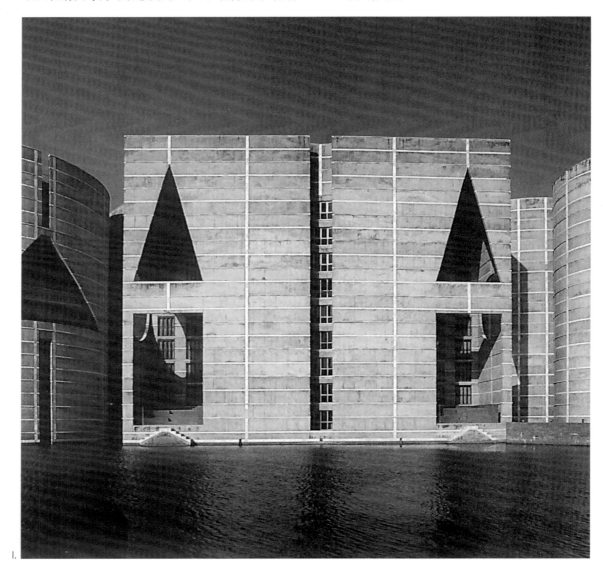

l. 办公部分
m. 中庭
n. 中庭的视野

l. Office section
m. Atrium
n. View of the atrium

light shaft with transparent material. As a result, the light space here has lost much of its original architectural force.

A total of 128 glass blocks are installed in the square concrete ribs, resulting in the diffusion of sunlight and the even illumination of the office rooms.

The horizontal strips in the masonry mark the level of the concrete cast forms, at heights of five feet (a story here is 10 feet, i.e., about 3 m). The joints between the individual concrete pouring strips are inlaid with white marble. In the external facade, the height of the floor slabs is visualized by strips provided as water-drip molds.

Within the purely centrally symmetrical configuration of the Assembly Hall, the geometry of the mosque has been slightly revolved out of the north-south axis, in order to orient it exactly toward Mecca. The cubic interior is flanked by four "light cylinders."

The Prayer Room is characterized by absolute introversion, with no view of the outside: no windows, no openings, no loopholes, nothing except light falling from above. Seen from the outside, the towers with the entrance gate indeed call to mind a citadel. Considered from the inside, however, the interior design of the Prayer Room proves exceedingly complex and extraordinary indeed. Semicircles and full circles setUnlike the offices, the light shafts here open to the sky. Wind and weather can therefore be experienced inside the mosque, and rays of sunlight variously enter the

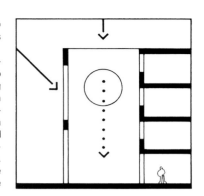

m.

n.

177

"在议会堂，我在平面内部引入了一个发光构件。如果你看到一列柱子，你可以说选择柱子就是选择了光。柱子就是框定了有光空间的实体。现在，反过来想想，想象柱子是中空的，也大得多，它们四周的墙可以发光。那么，空的地方就是房间，而柱子是光的给予者，可以呈现更复杂的形状，支撑着空间，给空间以光。我致力于将建筑元素发展成一个诗意的实体，在组合中超越自己的位置拥有自己的美。这样，它就类似于上面我提到的实心柱子，成为光的给予者。"（36）

值得注意的是，康用了许多年去寻找合适议会堂屋顶的解决方案。他最初的提议由于经过技术计算证明不可行，抑或仅仅是因为太贵了。直到 1972 年，即签订合同十年后，他才与工程师亨利·帕尔姆鲍姆（Harry Palmbaum）合作找到了解决方案。最终的屋顶设计在康的作品中是全新而且惊人的：一个伞形（双曲抛物面）横跨在八边形的议会堂之上，支撑点看起来非常脆弱。康用结构概念描述了一种新的方法：双曲抛物面的八个支撑作用的"伞尖"精确地在最大荷载点逐渐变细。然而，这种不合逻辑的向下

o.

o. 南立面
p. 祈祷室
q. 采光井

o. South facade
p. Prayer room
q. Light shaft

Prayer Room, depending on the time of day. Passing clouds strongly influence the character of the interior. Artificial light fixtures are hung low, in accordance with mosque tradition, to create a subdued, more intimate atmosphere during the night. The foyer beneath the Prayer Room is also illuminated by the four "light cylinders."

The very heart of the design for the National Capital is the Assembly Hall, the parliament chamber for the national legislative body, surrounded by its ring of "serving spaces": offices, ministers' offices, reception rooms, and TV and radio studios. The plan is based on an octagon covered by an umbrella-formed, vaulted roof. In 1964, Kahn wrote:

"In the Assembly I have introduced a light-giving element to the interior of the plan. If you see a series of columns, you can say that the choice of columns is a choice in light. The columns as solids frame the spaces of light. Now, think of it just in reverse, and think that the columns are hollow and much bigger and that their walls can themselves give light. Then, the voids are rooms, and the column is the maker of light and can take on complex shapes and be the supporter of spaces and give light to spaces. I am working to develop the element to such an extent that it becomes a poetic entity which has its own beauty outside of its place in the composition. In this way, it becomes analogous to the solid column I mentioned above as a giver of light." (36)

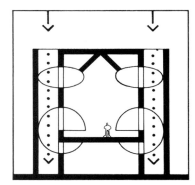

p.

q.

逐渐变细的伞尖不仅是一个结构工程问题：也给安装在内部的排水管带来了困难。拱顶上的大理石带纯粹是装饰性的：此处康引用了立面构造中的一个结构要素。

双曲抛物面屋顶令人想起文艺复兴建筑。然而在议会堂的屋顶，康似乎没有完全成功地借鉴历史建筑形式并将其诠释为一个全新的形式。议会堂建筑群复古、笨重、城堡般的建筑语言似乎与拱顶脆弱的外观形成了鲜明对比。

在达卡，一个"光环"围绕着飘浮的抛物面穹顶。光线从侧墙穿过抛物线的洞口进入室内。在穹顶之上，一个向内的带形窗刻入八边形内，玻璃表面与穹顶的边缘齐平。为了给议会堂提供更多的采光，康原本在"光环"外墙设置了额外的垂直槽形窗。然而由于眩光，现在这些窗户都用白色织物带遮住了。

站在议会堂中，真的会被他的大胆勇气和建筑的壮丽所感染。议会堂建筑群的确是路易斯·I. 康最迷人、最美丽的作品，其特质完全源自对光线的调节。他在达卡的成就是我们这个时代的建筑瑰宝。

"议会大厦的建筑形象源于这样一种理念，即保持一

It is noteworthy that Kahn worked for many years to find a suitable solution for the roof of the Assembly Hall. His initial proposals proved to be unfeasible on the basis of engineering calculations, or they were simply too expensive. It was only in 1972, i.e., ten years after award of contract that he found a solution, in collaboration with the engineer Palmbaum. The final roof design is new and surprising in Kahn's work: an umbrella form (hyperbolic paraboloid) which spans the octagon of the Assembly Hall on extremely fragile-seeming support points. Kahn describes a new approach with structural concepts: the eight supporting "umbrella tips" of the hyperbolic paraboloid taper precisely at points of greatest loads. It is, however, not only these illogically downward-tapering umbrella tips which represented a structural-engineering problem: the drain pipes installed on the inside also posed difficulties. The marble strips in the vaulted roof are purely decorative: here, Kahn plays with an element which is a structural component in the facade construction.

The hyperbolic-paraboloid roof calls Renaissance buildings to mind. In the Assembly Hall roof, however, Kahn appears not to have succeeded entirely in his borrowing of solutions from architectural history forms which he subjected to a completely new form of interpretation. The archaic, ponderous, castle-like language of the Assembly complex seems to stand in contradiction to the fragile appearance of the vaulted roof.

s.

r. 北立面
s. 议会堂

r. **North facade**
s. **Assembly hall**

种强烈的基本形式，给不同的内部需求以特定的形状，并从外观上将其表现出来。其形象是由混凝土和大理石建成的多面体宝石。"（37）

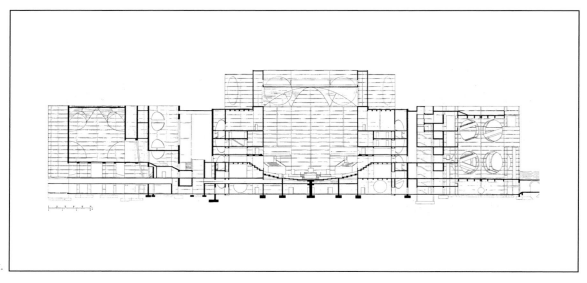
t.

u.

In Dhaka, a "ring of light" encircles the intrinsically floating paraboloid dome. The light is admitted from the side walls, through the parabola openings, into the interior. Above the dome, an inwardly oriented strip window is inscribed in the octagon, and the glazed surface is flush with the border of the dome. To provide more light for the Hall, Kahn originally installed additional, vertically slotted windows on the outside wall of the "ring of light." Owing to glare, however, they are now covered with white fabric strips.

Standing in the Assembly Hall, one is truly overwhelmed by the daring and by the magnificence of the architecture. The Assembly complex is indeed Louis I. Kahn's most fascinating and most beautiful work characterized as it utterly is by the idea of modulation of light. His achievements in Dhaka represent an architectural jewel of our age.

"The architectural image of the Assembly building grows out of the conception to hold a strong essential form to give particular shape to the varying interior needs, expressing them on the exterior. The image is that of a many-faceted, precious stone, constructed in concrete and marble." (37)

v.

t. 剖面图
u. 抛物面屋顶
v. 议会堂

t. Section
u. Paraboloid roof
v. Assembly hall

w. 采光空间
x. 屋顶
y. 抛物面屋顶

w. Light space
x. View of the roof
y. Paraboloid roof

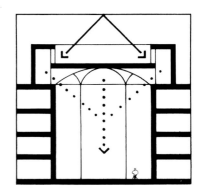

光线控制类型

按年代顺序排列的项目列表追溯了康在光线控制和调节方面的发展历程。从 1924 年的学生作品开始到 1971—1974 年的伯克利图书馆为止，这些项目展示了康在光与空间的设计中相当广泛的设计手法。康在每一个建筑挑战中为了找到清晰的答案所做出的努力，仍是赋予他所有作品特质的基本原则。康没有落入公认的原则或法则之中，而是将每个设计都视为从头开始的全新挑战。对他来说，对光线的控制和空间的构建是不可分割地交织在一起的。

在他早期的作品中，康将窗户作为普通的单层墙面上的一个简单洞口。遮阳板和遮阳屋顶顺理成章地出现在康的发展过程中，努力更好地控制直射阳光。垂直滑动木质嵌板的概念曾多年主导着康的理念，但后来被摒弃掉了——和他的水平遮阳板一样——转而开始采用独特的创新解决方案。双层墙体是康自己独创的解决光线控制和调节问题的方式。他从未停止对这些"光过滤器"的发展和完善。于是室内和室外不再像他早期作品中那样被简单的带有洞口的墙体分割：界面变成了多层空间，调节着光线。康将这些垂直的光过滤器放在主立面前面，或是将二

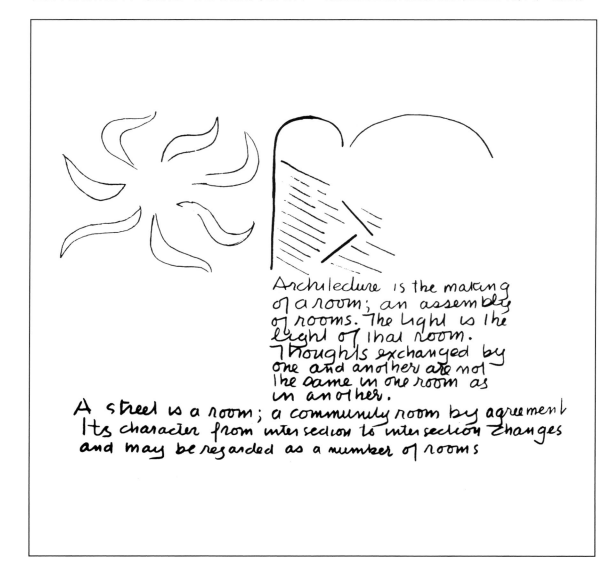

路易斯·I. 康的草图
Sketch by Louis I. Kahn

The chronological listing of projects traces Kahn's development process in light control and modulation. Projects beginning with student work from 1924 and ending with the Berkeley library from 1971-1974 reveal an extraordinarily broad spectrum of solutions for the design of light and space. Kahn's struggle to find the coherent answer for each architectural challenge remains the basic principle characterizing his complete work. Kahn did not fall back on received principles or rules, but saw each design as a challenge to start anew, from scratch. For him, the control of light and the structuring of space are indivisibly interwoven.

In his early work, Kahn installed the window as a simple opening in an ordinary single-shell wall. The brisesoleil and the canopy roof followed as logical steps in Kahn's development process, in efforts to better control direct sunlight. The concept of vertical sliding wooden panels occupied Kahn's attention for years, but was later discarded – as were his horizontal brisesoleils – in favor of distinctively innovative solutions. Double-shell walls offered Kahn the possibility of solving the problem of light control and modulation in his own, original manner. Kahn never ceased to further develop and to sophisticate these "light filters." Indoors and outdoors were then no longer divided by a simple wall with openings, as in his early work: the interface became multi-layer space in which light is modulated. Kahn placed these vertical light filters in front of the main facades, or

TYPOLOGY OF LIGHT CONTROL

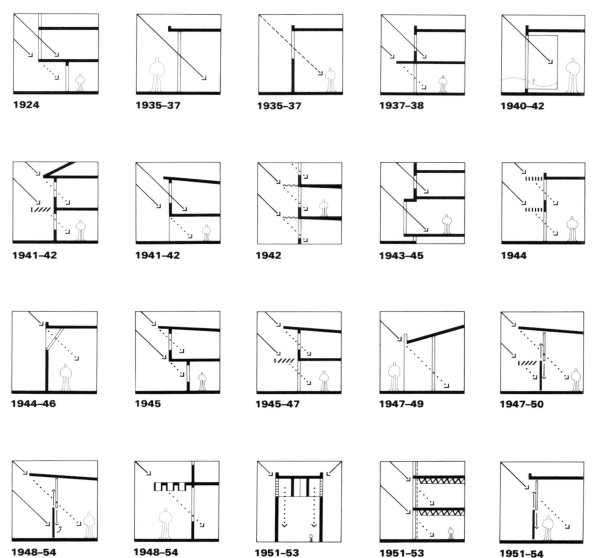

者整合在一起；在他晚期的作品中，这些光过滤器越来越多地采用古老的形式。带有圆形、三角形或是矩形洞口的"遮光罩"赋予康的建筑真正与众不同的特质。这些空间不仅能够调节光线，还能提供通风和眺望室外的视野。尽管康开始没有采用天窗采光，但在他晚期的作品中，天窗作为光线控制元素变得越来越重要。从金字塔形屋顶上的简单洞口，一直到复杂的拱顶结构，康为中央和外围房间的天窗采光发展出了一系列惊人的解决方案。它们在这张类型图中按项目时间顺序排列。康对设计问题最佳解决方案的不懈探索在我的研究中一直延伸到1971年，仅在他去世前三年。一些1971—1974年间的未建成项目，从我们可获得的图纸来看，其差异并不明显。

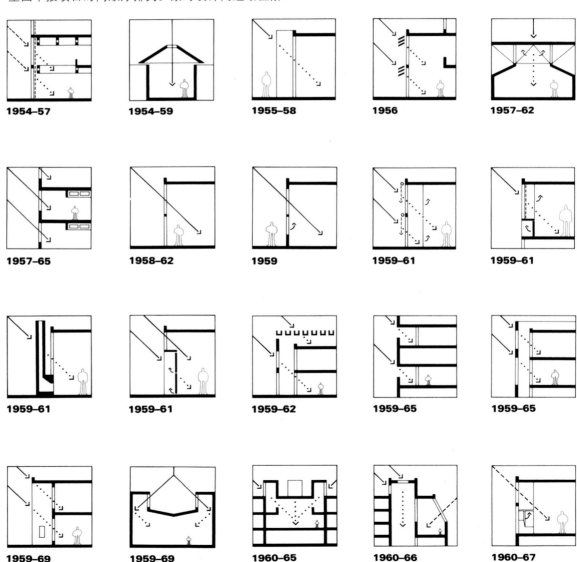

integrated the two elements; in his late work, these light filters increasingly took on archaic forms. "Light shields" featuring circular, triangular, or rectangular openings endow Kahn's architecture with its truly distinguishing traits. These interstices enable not only light modulation, but also provide ventilation and framed views toward the outside. Although Kahn did not employ the skylight as an element of illumination at the beginning, it became increasingly important for light control in his late works. Ranging from simple openings in a pyramidal roof, all the way to a complex dome structure, Kahn developed an astonishing spectrum of architectural solutions for skylight illumination of central as well as peripheral rooms. They have been arranged in this typology by chronological order according to the year of project award. Kahn's never-ending search for the optimal solution to a design problem is reflected in the fact that my analytical study extends to the year 1971, only three years before his death. A number of unbuilt projects follow during the period of 1971 to 1974, but their distinguishing attributes are not evident from the schematic representations available to us.

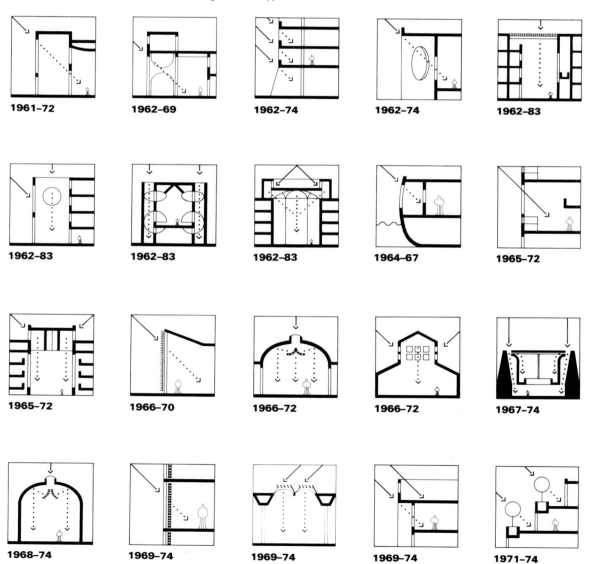

参考文献

REFERENCES

1. Wurman, R.S.; Feldman, E.: The notebooks and drawings of Louis I. Kahn. (MIT Press, Cambridge, Mass., and London, England, 1973).
2. Wurman, R.S.: What will be has always been. The words of Louis I. Kahn; from a conversation with Robert Wemischner, April 17, 1971; pp. 121 (Access Press / Rizzoli, New York, 1986).
3. Wurman, R.S.: What will be has always been. The words of Louis I. Kahn; the white light and the black shadow, lecture at Princeton University, March 6, 1968; pp. 15 (Access Press / Rizzoli, New York, 1986).
4. Wurman, R.S.: What will be has always been. The words of Louis I. Kahn; from a conversation with Jaime Mehta, October 22, 1973; pp. 230 (Access Press / Rizzoli, New York, 1986).
5. Wurman, R.S.: What will be has always been. The words of Louis I. Kahn; the invisible city – International Design Conference at Aspen, June 19, 1972; pp. 153 (Access Press / Rizzoli, New York, 1986).
6. Wurman, R.S.: What will be has always been. The words of Louis I. Kahn; architecture and human agreement, a Tiffany lecture, October 10, 1973; pp. 215 (Access Press / Rizzoli, New York, 1986).
7. Wurman, R.S.: What will be has always been. The words of Louis I. Kahn; University of Cincinnati, College of Design, Architecture, Art and Planning; pp. 75 (Access Press / Rizzoli, New York, 1986).
8. Wurman, R.S.: What will be has always been. The words of Louis I. Kahn; the invisible city – International Design Conference at Aspen, June 19, 1972; pp. 151 (Access Press / Rizzoli, New York, 1986).
9. Wurman, R.S.; Feldman, E.: The notebooks and drawings of Louis I. Kahn. (MIT Press, Cambridge, Mass., and London, England, 1973).
10. Wurman, R.S.; Feldman, E.: The notebooks and drawings of Louis I. Kahn. (MIT Press, Cambridge, Mass., and London, England, 1973).
11. Wurman, R.S.: What will be has always been. The words of Louis I. Kahn; progressive architecture 1969, special edition, wanting to be: the Philadelphia school; pp. 89 (MIT Press, Cambridge, Mass., and London, England, 1973).
12. Wurman, R.S.: What will be has always been. The words of Louis I. Kahn; Perspecta 3, the Yale architectural journal 1955; pp. 2 (MIT Press, Cambridge, Mass., and London, England ,1973).
13. Wurman, R.S.: What will be has always been. The words of Louis I. Kahn; lecture, Drexel Architectural Society, Philadelphia, Pa., November 5, 1968, pp. 29 (MIT Press, Cambridge, Mass., and London, England, 1973).
14. Ronner, H.; Jhaveri, S.: Complete Work 1935 – 74; 2nd ed., pp. 65 (Birkhäuser Verlag, Basel, 1987).
15. Ronner, H.; Jhaveri, S.: Complete Work 1935 – 74; 2nd ed., pp. 83 (Birkhäuser Verlag, Basel, 1987).
16. Ronner, H.; Jhaveri, S.: Complete Work 1935 – 74; 2nd ed., pp. 122 (Birkhäuser Verlag, Basel, 1987).
17. Ronner, H.; Jhaveri, S.: Complete Work 1935 – 74; 2nd ed., pp. 120 (Birkhäuser Verlag, Basel, 1987).
18. Wurman, R.S.: What will be has always been. The words of Louis I. Kahn; voice of America; pp. 8 (Access Press / Rizzoli, New York, 1986).
19. Ronner, H.; Jhaveri, S.: Complete Work 1935 – 74; 2nd ed., pp. 128 (Birkhäuser Verlag, Basel, 1987).
20. Ronner, H.; Jhaveri, S.: Complete Work 1935 – 74; 2nd ed., pp. 134 (Birkhäuser Verlag, Basel, 1987).
21. Ronner, H.; Jhaveri, S.: Complete Work 1935 – 74; 2nd ed., pp. 344 (Birkhäuser Verlag, Basel, 1987).
22. Ronner, H.; Jhaveri, S.: Complete Work 1935 – 74; 2nd ed., pp. 189 (Birkhäuser Verlag, Basel, 1987).
23. Ronner, H.; Jhaveri, S.: Complete Work 1935 – 74; 2nd ed., pp. 208 (Birkhäuser Verlag, Basel, 1987).
24. Ronner, H.; Jhaveri, S.: Complete Work 1935 – 74; 2nd ed., pp. 222 (Birkhäuser Verlag, Basel, 1987).
25. Ronner, H.; Jhaveri, S.: Complete Work 1935 – 74; 2nd ed., pp. 293 (Birkhäuser Verlag, Basel, 1987).
26. Wurman, R.S.: What will be has always been. The words of Louis I. Kahn; comments on the library, Phillips Exeter Academy, Exeter, 1972; pp. 179 (Access Press / Rizzoli, New York, 1986).
27. Wurman, R.S.: What will be has always been. The words of Louis I. Kahn; comments on the library, Phillips Exeter Academy, Exeter, 1972; pp. 180 (Access Press / Rizzoli, New York, 1986).
28. Wurman, R.S.: What will be has always been. The words of Louis I. Kahn; comments on the library, Phillips Exeter Academy, Exeter, 1972; pp. 130 (Access Press / Rizzoli, New York, 1986).
29. Louis I. Kahn and the Kimbell Art Museum: Light is the theme; comments on architecture by Louis I. Kahn; pp. 33 (Kimbell Art Foundation, Fort Worth, Tex., 1975).
30. Ronner, H.; Jhaveri, S.: Complete Work 1935 – 74; 2nd ed., pp. 345 (Birkhäuser Verlag, Basel, 1987).
31. Wurman, R.S.: What will be has always been. The words of Louis I. Kahn; comments on the library, Phillips Exeter Academy, Exeter, 1972; pp. 216 (Access Press / Rizzoli, New York, 1986).
32. Ronner, H.; Jhaveri, S.: Complete Work 1935 – 74; 2nd ed., pp. 365 (Birkhäuser Verlag, Basel, 1987).
33. Ronner, H.; Jhaveri, S.: Complete Work 1935 – 74; 2nd ed., pp. 363 (Birkhäuser Verlag, Basel, 1987).
34. Ronner, H.; Jhaveri, S.: Complete Work 1935 – 74; 2nd ed., pp. 234 (Birkhäuser Verlag, Basel, 1987).
35. Ronner, H.; Jhaveri, S.: Complete Work 1935 – 74; 2nd ed., pp. 238 (Birkhäuser Verlag, Basel, 1987).
36. Ronner, H.; Jhaveri, S.: Complete Work 1935 – 74; 2nd ed., pp. 239 (Birkhäuser Verlag, Basel, 1987).
37. Ronner, H.; Jhaveri, S.: Complete Work 1935 – 74; 2nd ed., pp. 257 (Birkhäuser Verlag, Basel, 1987).

图片来源

FIGURE CREDITS

The Architectural Archives, University of Pennsylvania, Philadelphia: figs. 84 b (Meyers Col.), 87 ar b (Meyers Col.), 92 bl (Komendant Col.), 134 (Wilcots Col.), 157 l (Meyers Col.)

"L'Architecture d'Aujourd'hui", Paris, January 1963: figs. 83 br, 113 ar br, 127

Linda Brenner / William Christensen, Philadelphia, Scale Models and Molds: figs. 112 br

Urs Büttiker, Basel: Photo on book cover, figs. 25, 27, 29, 31, 33, 35, 37,39 ar bl br, 41, 42 bl ar br, 43, 44, 45, 47, 48 ar, 49, 51, 52 a b, 53 al ar bl br, 54 bl, 55, 57 l br, 59, 62 bl, 63, 65, 66 al br, 67, 68 b, 69 a, 70 bl, 71 l br, 72 bl, 73, 74 bl c, 78 br, 79 a bl br, 81, 84 a, 85 l br, 90 bl r, 91, 93, 94 b, 95, 100, 101, 103 l br, 104, 105 a b, 106 bl br, 107 l br, 110 bl ar,111, 114 a b, 115, 117, 118 ar, 119 l br, 122 bl r, 123 l br, 128, 129 l, 130 bl r, 131, 132 bl, 133, 135 l br, 136 r, 138 bl r, 139 l br, 140 b, 141, 146, 147, 148 r, 149 l br, 150 bl, 154 ar, 155 l br, 157 r, 159 l, 162, 163 l br, 164 bl br, 165, 166 ar br, 167, 169, 171 l br, 172 al ar bl br, 173, 174, 175, 176, 177 l br, 178, 179 l, 180,181, 182 b, 183, 184 bl r, 185, 187,188, 189

Christian Dill, Basel: fig. 102 ar

John Ebstel© Keith de Lellis Gallery, New York: figs. 75, 82, 83 l

Eugen Eisenhut, Zürich / Esther Zumsteg, Basel: figs. 151 al ar bl br, 152 a b, 153

Government of the Peoples Republic of Bangladesh, Dhaka, Department of Architecture and Public Works, "The National Assembly Building, 1984": figs. 168, 170

Louis I. Kahn Collection, University of Pennsylvania and Pennsylvania Historical and Museum Commission, Philadelphia (housed in the Harvey & Irwin Kroiz Gallery, the resources of the Kahn Collection are published with the permission of The Architectural Archives, University of Pennsylvania): figs. 11, 15, 21, 39 al, 40, 46 al ar bl br, 48 bl br, 50 al ar bl br, 54 r, 56, 58 a b, 60, 61, 62 r, 64 al ar bl br, 66 ar, 68 a, 69 b, 70 r, 72 ar br, 74 r, 76, 78 a, 80 al ar b, 86, 87 al ar b, 88 a bl br, 89 a b, 92 a br, 94 al ar, 96 bl r, 97 al ar b, 102 bl br, 106 a, 108 al ar b, 109 l br, 110 br, 112 ar, 113 al bl, 118 bl br, 120 r, 121 l br, 124 a b, 125 a b, 126, 129 br, 132 ar br, 136 bl, 137, 140 a, 143 br, 144, 145 l br, 148 bl, 150 ar br, 154 bl al br, 158, 160 a br, 164 al ar, 166 bl, 182 a, 186

Kimbell Art Museum, Fort Worth: figs. 142, 143 l

Hans H. Münchhalfen, Basel: Photo of the portrait

Walter Kläy, Basel: figs. 77, 78 bl, 98

Officina Edizioni, Roma, Florindo Fusaro, "Il Parlamento e la nuova Capitale a Dacca di Louis I. Kahn, 1962-74", 1985: figs. 179 br

Philadelphia Museum of Art, Philadelphia: fig. 19
Louis I. Kahn, "Architecture Comes from the Making of a Room..." (For City / 2 Exhibi-tion) 1971, Charcoal on yellow tracing paper, Sheet: 33 ¾ x 33 ¾ inches (85.7x 85.7 cm), Gift of the Artist, 1972, 1972-32-4

Verlag für Architektur Artemis Zürich und München, 1978: figs. 112 bl, 116, 120 bl

Gunnar Volkmann, Leipzig: figs. 159 br, 160 bl, 161

William Whitaker, Philadelphia: fig. 99

Yale Center for British Art, New Haven: fig. 156
Exterior View, Photo Credit: Richard Caspole, Yale Center for British Art

Esther Zumsteg, Basel: fig. 7

Every effort has been made to credit the sources and photographers of all illustrations in this book; if there are any errors or omissions, please contact us so that corrections can be made in any subsequent edition.

第一版致谢

我想在此对所有为这本书提供过帮助的人们表示由衷感谢。没有这么多人给予我的鼓励和支持，今天就没有这本书的出版。

宾夕法尼亚大学建筑档案馆的茱莉亚·摩尔·康弗斯女士以其言行为我提供了帮助。我要特别感谢沃纳·布拉瑟，是他邀请我到巴塞尔艺术馆听了一场关于路易斯·I. 康的讲座。正是这场讲座给我启发，促使我将收集整理的资料以图书形式出版。在本书长时间的撰写过程中，伯克豪舍出版社的汉斯—彼得·托尔慷慨地给我时间、理解和信任。柏林的玛蒂娜·杜特曼在设计和文本编辑方面为本书提供了帮助。大卫·比恩将文本译为英文。海因茨·希尔特布鲁纳为本书设计了封面。

还要衷心感谢艾斯特·祖姆斯蒂格和汉斯—约格·查尼，他们仔细严谨地帮助修改排版，并绘制了图示和技术图纸。

在美国、印度、孟加拉和瑞士，我遇到了许多慷慨提供路易斯·I. 康以及与他的建筑相关资料的人们。在此我还要向他们所有人表示感谢。

CREDITS FOR THE FIRST EDITION

I would like to extend my heartfelt appreciation here to all of those who helped me with this book. Its publication would not have been possible without the encouraging support which I received from many people.

Mrs. Julia Moore Converse at the Architectural Archives of the University of Pennsylvania provided help in the form of word and deed. I am particularly grateful to Werner Blaser, who invited me to a lecture on Louis I. Kahn at the Kunsthalle in Basel. It was this lecture which inspired me to compile my collected information in book form. Hans-Peter Thür of Birkhäuser Publishers generously granted me time, understanding, and trust during the long effort with my book. Without the support of Martina Düttmann in Berlin, who collaborated on the layout and edited the text, it would not have been possible to finish the book. David Bean translated the text into English. Heinz Hiltbrunner designed the cover.

Heartfelt thanks also go to Esther Zumsteg and to Hans-Jörg Tschäni, who both enthusiastically and carefully helped to revise the layout, and who produced the emblems and the technical drawings.

In the USA, India, Bangladesh, and here in Switzerland, I met many people who have generously provided information on Louis I. Kahn and his architecture. I would also like to extend my thanks to all of them at this point.

中 / 英文对照版致谢

我的《路易斯·I. 康：光与空间》一书的德 / 英文对照版于 1993 年首次出版，之后出版了英 / 日文对照版。现在又出版了中 / 英文对照版。

天津大学仁爱学院卢紫荫老师和天津大学出版社刘大馨编辑提议并促成了本书的出版。他们想要为学生和对路易斯·I. 康迷人的建筑感兴趣的人们重新出版本书，这令我十分开心，尤其是在本书第一版已经绝版的情况下。我要衷心感谢刘大馨先生卓有成效的"洲际合作"——当然，也要感谢卢紫荫女士将其翻译为中文。

本书的出版还要感谢费城宾夕法尼亚大学建筑档案馆馆藏主任威廉·惠特克先生，没有他的帮助本书是无法出版的。他以极大的精力投入和渊博的专业知识支持了这个图书项目。非常开心能和他一起工作。

汉斯—彼得·托尔先生是伯克豪舍出版社的前出版经理，也是我这本书的第一位编辑，他为本书的出版提供了法律方面的建议，对我有极大的帮助。

我要感谢我的太太伊丽莎白·阿达杜拉布·布提克的支持。她的意见和判断总是给我启发与激励。

为了减少高昂的版权费，博物馆、出版社、期刊、朋友和专业的同事给我免费提供了许多草图、图片和照片。没有他们慷慨的帮助，这本书也是不可能出版的。我要特别感谢琳达·布伦纳、威廉·克里斯滕森、艾丽西亚·卡纳莱斯、克里斯蒂安·迪尔、尤金·艾森胡特、沃尔特·克莱、汉斯·H. 慕奇哈芬以及艾斯特·祖姆斯蒂格。

我们希望这一版本使对"光与空间"的探索再次受到关注。希望路易斯·I. 康"永恒的"建筑启迪更多的读者。

CREDITS FOR THE CHINESE/ENGLISH EDITION

My book "Louis I. Kahn: Light and Space" in English/German was published for the first time in 1993, followed by an American and a Japanese version. Now it is also available in a Chinese/English edition.

This was made possible by the request and the initiative of Teacher Lu Ziyin of Tianjin University Renai College and Editor Liu Daxin from the Tianjin University Press. His idea to reissue the book, for students and persons interested in Louis I. Kahn's fascinating and impressive architecture, elated me from the beginning, especially since the first editions are out of print. My heartfelt thanks go to Mr. Liu Daxin for the fruitful "transcontinental" cooperation - and of course, also to Ms. Lu Ziyin for translating the text into Chinese.

This publication would not have been possible without the assistance of the Collection Manager Mr. William Whitaker from the Architectural Archives of the University of Pennsylvania in Philadelphia. He supported the book project with great commitment and his immense knowledge. It was a joy to work with him.

Mr. Hans-Peter Thür, the former Publishing and Managing Director of Birkhäuser Publishers and the first editor of my book, has given me legal advices concerning this publication. His references were of great help to me.

I would like to thank my wife Elisabeth Abd'Rabbou Büttiker for her support. Her opinion and judgment are always an inspiration to me.

Museums, publishers, journals, friends and professional colleagues have made many sketches, pictures and photos available free of charge in order to avoid high copyright fees. The publication of the book would not have been possible without their generous help. I am particularly grateful to Linda Brenner, William Christensen, Alicia Canales, Christian Dill, Eugen Eisenhut, Walter Kläy, Hans H. Münchhalfen and Esther Zumsteg.

We hope and we wish that this edition will make the explored theme "Light and Space" accessible again. May the "timeless" architecture by Louis I. Kahn inspire and enlighten our readers.